5/26/92

D1207127

Multivariate Analysis

Other statistical books by Sir Maurice Kendall

Time-Series
Rank correlation methods
Exercises in theoretical statistics
A course in the geometry of *n* dimensions[*]

With Professor A. Stuart
The advanced theory of statistics (3 volumes)

With G. Udny Yule
An introduction to the theory of statistics

With Professor P. A. P. Moran
Geometrical probability[*]

WORKS EDITED BY SIR MAURICE KENDALL
Mathematical model building in economics and industry (2 volumes)

With Professor A. Stuart
Statistical papers of George Udny Yule

With Professor E. S. Pearson
Studies in the history of statistics and probability: Vol. 1

With Professor R. L. Plackett
Studies in the history of statistics and probability: Vol. 2

With Professors F. N. David and D. E. Barton
Symmetric functions and allied tables

[*] A volume in "Griffin Statistical Monographs and Courses"
Full statistical list available from Charles Griffin & Co. Ltd.

Multivariate Analysis

Sir MAURICE KENDALL, Sc.D., F.B.A.

Second edition

CHARLES GRIFFIN & COMPANY LTD
London and High Wycombe

CHARLES GRIFFIN & COMPANY LIMITED
Registered Office:
Charles Griffin House, Crendon Street, High Wycombe,
Bucks HP13 6LE

Copyright © Sir Maurice Kendall, 1980

First published   1975
Second edition   1980

ISBN: 85264 264 4

Set by Alden Press Ltd, Oxford, London and Northampton
Printed in Great Britain by J. W. Arrowsmith Ltd, Bristol

Preface to First Edition

It is now more than fifteen years since I published a monograph entitled *A Course in Multivariate Analysis*. To judge from the sales, it has continued to supply a need and has passed through several printings without much alteration. However, in the meantime we have had the computer revolution which has, perhaps, affected the development of multivariate analysis more than any other branch of statistics. I therefore decided to re-write the work completely and to extend it into a book. The methods which are now available for the exploration of multivariate situations have turned the present volume into what is, essentially, a new work. But I have adhered to my original plan in that this book is intended for the practitioner, not the theorist, although a good deal of theory is unavoidable.

I am indebted to many colleagues for discussions and illustrative material, notably to Professor Martin Beale of Scientific Control Systems Limited, Mr Peter Herne of the National Coal Board, Dr Morton Brown of the University of Los Angeles, and to Dr Vijay Verma and Dr Kenneth Williams. I think that some of them take a different view from mine on one or other sections of the treatment, but in a growing subject like this, which is still as much an art as a science, that was to be expected and in no way diminishes my gratitude to them. I am especially grateful to Mr E.V. Burke of Charles Griffin & Co. for his care in steering through the press some rather difficult manuscript.

As always I should be glad to be notified of any misprints, obscurities or omissions.

M.G.K.

London
April, 1975

Preface to Second Edition

The author is most gratified that the steady and continuing demand for this book has given cause for a further edition. The opportunity has been taken to make some minor improvements to the text and to correct a few errors in the script. References have been added to recent work in this field. I am grateful to Professor J. Durbin, of the London School of Economics, for the advice he has given during the preparation of this edition.

London   M. G. K.

1980

Contents

1

Introduction

1.1   The characteristic feature of multivariate analysis is the consideration of a set of n objects, on each of which are observed the values of p variables. The set of objects may be complete or it may be a sample from a larger set. The variables may be continuous or discontinuous, and themselves may be a subset of a larger group. Given such a complex, we require to study it for a number of different purposes of which the following are the most important:

(1)   Structural simplification.   The object here is to "see the wood for the trees" by examining whether, by variate-transformations or otherwise, there are simpler ways of representing the complex under study, e.g. by transforming a set of interdependent variables to independence, or reducing the dimensionality of a complex.

(2)   Classification.   The question is whether the objects fall into groups or clusters, as against being more or less haphazardly scattered over the domain of variation.

(3)   Grouping of variables.   Whereas classification is concerned with the grouping of the objects, we may be interested in whether the variables also fall into cognate groups.

(4)   Analysis of interdependence.   The object is to examine the interdependence of variables, the possibilities ranging from independence to collinearity, i.e. a situation in which one variable is a linear function of others (or, more generally, is a nonlinear function of others).

(5)   Analysis of dependence.   In (4) the variables are regarded as all standing on an equal footing so far as concerns their mutual relationships. In dependence analysis one or more variables are singled out for the examination of their dependence on the others, as in regression analysis.

(6)   Hypothesis construction and hypothesis testing.

1.2   Formally, then, we may define multivariate analysis as the branch of statistics which is concerned with the relationships among sets of dependent variables and the individuals which bear them. However, the content of a

1

course on the subject is more easily defined by enumerating the topics to be covered rather than by formal definition. Some branches are covered in an earlier part of the student's training, for example, correlation and regression. Others are not even mentioned in quite advanced texts, for example, cluster analysis and the analysis of functional relationships. The present course will try to exhibit the subject as a connected whole, but greater emphasis will be placed on practical application than on mathematical refinement. The mathematics of the subject, fascinating and challenging as they are to the theorist, may make greater demands on the time of the practising statistician than he can allot to them.

1.3   There is one respect in which, by reason of historical development and current usage, the subject of multivariate analysis is restricted in scope: it does not usually consider the time-tracks of multivariate complexes. This part of statistics is usually reserved for the study of time-series themselves. The distinction, perhaps, is more convenient than logical and we shall occasionally have to consider temporal effects in some of our examples. The Compleat Statistician, of course, requires expertise in both multivariate analysis and time-series, but in this book we shall not be much concerned with the latter, on which a separate study has been published (Kendall, 1973).

1.4   To give concreteness to the exposition we may refer briefly to some of the types of practical problems with which multivariate analysis is concerned and some of the fields in which it is currently applied.

(1)   Agriculture.   A number of n different areas each produce yields of p different crops. Can we make any comparisons of general productivity between areas, and if so, how?

(2)   Anthropology.   For a set of Red Indian tribes there are recorded a number of items such as whether the tribe has a rain god, whether it uses totems, whether it has any agriculture. It is required from this material to decide whether a tribe belongs to certain ethnic groups, or to suggest what such groups might be.

(3)   Archaeology.   On the basis of funerary remains (especially pottery and jewellery) in a group of graves it is required to arrange them in temporal sequence, for example by considering differences in pattern or ornamentation.

(4)   Biometrics.   A number of skulls are dug up on an ancient burial ground. They may all come from one race or may be a mixture of two opposing races, friend and foe having been flung into one pit together. An unlimited number of measurements can be taken on any one skull. What are the best measurements to take, what is the minimum number we require, and how do we use them to test homogeneity or heterogeneity in the sample?

(5) Economics. Each year there is produced for a given country data which are in some way bound up with its general business activity, such as national income, rate of interest, freight-car loadings, unemployment and so forth. Can we produce an index-number of "business activity" from this complex, and if we can does it have any objective meaning?

(6) Education. A number n of candidates take an examination in p parts and are given a mark for each part. What is the best system of arranging the candidates in order of merit, and does the notion of "order of merit" have any justifiable meaning? And if we offer prizes for the best performances, what should be the ratio of the values of first and second prizes?

(7) Experimentation. By accident or design an experiment is conducted without proper balance in the factors and with multiple measurements on the outcome. How do we assess the significance of the main effect?

(8) Industry. A firm of tailors making ready-to-wear garments wishes to produce enough to cover the requirements of a large clientele with the minimum of misfits and unsold garments. Which measurements on the human frame are required, and how many of each type of garment should be produced?

(9) Medicine. A number of symptoms (presence or absence) are recorded on patients who may or may not have a certain disease. Which symptoms are most discriminatory, and what are the best ways of using the material for preliminary diagnosis?

(10) Meteorology. Records are available for rainfall, mean temperature, barometric pressure, etc., at various places and perhaps at various points of time. The problems might be to measure the relationships among variables, or among places or at various epochs, or all three together.

(11) Physics. A set of plastics are tested for various physical properties such as resilience, strength, ability to withstand high temperatures, resistance to corrosion. Can we detect in the results any systematic effects such as would enable us to predict them from the molecular constitution of the plastics? Can we then use the results to design better plastics for given purposes?

(12) Sociology. The replies given by members of a population to a questionnaire are expected to vary according to the position in which they see themselves in the social hierarchy. Information is collected about certain objective properties, such as possession of a telephone, type of education, rent paid, and so forth. Can an index of social class be constructed from such material, or is the concept essentially multidimensional?

These examples, taken more or less at random, illustrate the wide field of application of multivariate methods. Other examples will occur in the sequel.

1.5   As we shall see, some of the methods familiar in univariate analysis are capable of extension to multivariate situations; such, for example, are moment-statistics (means, variances, measures of skewness) and methods based on them (correlation, regression, analysis of variance). Some, however, are new (principal components, factor analysis, canonical correlations, classification). There are certain other features which render multivariate analysis much more than an extension to p dimensions of unidimensional techniques:

(1)   Many of the practical situations which confront us are not probabilistic in the ordinary sense. We may, for example, have the whole set of objects available for scrutiny, as if we are classifying British towns. Or our data may be a subset of existent objects but not a random sample, e.g. the patients who arrive for treatment at a hospital or the visible stars in the sky. It is a mistake to try and force the treatment of such data into a classical probabilistic mould, even though some subjective judgement in treatment and interpretation may be involved in the analysis of the results.

(2)   The development of the electronic computer has added enormous scope to multivariate methods in practice. Much of the computation which can now be done in a matter of minutes, or even seconds, was impossible a generation ago. This is doubtless an enormous advantage, but it is not without its adverse side. A statistician working with machine programs which he did not write is compelled to take on trust a good deal of what formerly he could check himself. Machine programs cannot be accepted in blind faith, and the experienced statistician who has to rely on software written by others would do well to have in his armoury a set of test problems with which he can satisfy himself that the machine is producing the kind of accuracy that he requires.

(3)   Even when the sample we have is a random one, we may encounter difficulty in analysing it because of non-normality in the parent distribution. In univariate statistics we have a whole range of frequency distributions for study (the discontinuous group such as Poisson and hypergeometric, or continuous ones such as Pearson curves or Edgeworth expansions). With their help the effect of non-normality on procedures based on normal variations can be examined in some detail. For more than one dimension the specification of a flexible and realistic set of surfaces is much more difficult. Most theoretical work in multivariate statistics is based on the assumption that the parent population is multinormal, and its robustness under departures from normality is very often difficult to determine with any exactitude.

(4)   The pictorial representation of data is simple enough in two dimensions and, with some trouble, can be carried out for three. For more than three it is virtually impossible and the visual presentation of a scatter of n points in p dimensions can only be handled by projecting it in various ways on to planes. There are in existence some sophisticated machine programs which will display on a video screen the projection of the scatter on to an arbitrary plane and enable us, so to speak, to look all round the scatter from many different points of view. Sometimes a three-dimensional picture can be suggested by the use of colour in two dimensions. These devices are a help, but are not a complete solution to the problem of visualizing the internal structure of the point-complex. Anyone who has tried to form a mental picture of an engineering component from the two-dimensional drawings of plan, front and side elevation will appreciate the difficulty.

(5)   In univariate statistics it is often possible to devise methods which are distribution-free and therefore gain in generality. They all depend on properties of order. In more than one dimension there is no orderability in the linear sense, and although a few devices are available, especially for two-dimensional material, to evade the problem, there are few distribution-free multivariate methods. This is particularly unfortunate in view of the remark in the foregoing paragraph (3) concerning the deficiency of families of multivariate distributions which can be fitted to data.

1.6   Although multivariate methods are many and various, there underlies nearly all of them a desire to simplify and to reduce the complexity of the problem. Theoretical and practical reasons all point in the same direction. For example, on the practical side we may wish to reduce the number of variables simply to save computational effort; or we may wish to avoid variables which are expensive to observe, or involve a lot of delay in measurement, provided that nothing serious is lost in the purpose of the inquiry. On the theoretical side we may wish to reduce the dimension number, even at the expense of sacrificing some information; or we may wish to transform so as to get rid of a number of nuisance parameters.

Example 1.1

Consider, for example, a case in which we have p variables and are interested in their variances and correlations. There are p means, p variances and $\frac{1}{2}p(p-1)$ correlations to be considered, $\frac{1}{2}p(p+3)$ parameters altogether. In the univariate case there are only two; in the bivariate case five; for $p = 10$ there are no fewer than 65. Clearly if we can transform the variables to a non-correlated set we have very substantially reduced the complexity of the representation, $\frac{1}{2}p(p-1)$ parameters then vanishing. Or again, if we reduce

the dimensionality from p to $(p-1)$ we get rid of $\frac{1}{2}p(p+3) -$
$\frac{1}{2}(p-1)(p+2) = p+1$ parameters.

Admittedly, circumstances arise where we should not wish to specify the system in great detail by estimating every one of the parameters. It will nevertheless be clear that the more we can reduce their number the closer we are to a more comprehensible model of the system structure.

1.7 For the time being we shall suppose that our variables are all continuous. The typical data can then be arranged in a $p \times n$ matrix:

$$
\begin{array}{llll}
x_{11} & x_{12} & \ldots \ldots & x_{1n} \\[6pt]
x_{21} & x_{22} & \ldots \ldots & x_{2n} \\[6pt]
x_{p1} & x_{p2} & \ldots \ldots & x_{pn}
\end{array}
\tag{1.1}
$$

Here x_{ij} is the value of the ith variable borne by the jth member. In actual practice, since n is usually much larger than p, it is often convenient to print the data with p running along the top of the page and n running down.

It will usually be convenient to measure each variable about the mean of its n values. We represent this by a subscript period, thus:

$$
x_{i.} = \frac{1}{n} \sum_{j=1}^{n} x_{ij}
\tag{1.2}
$$

If we postmultiply the matrix of values so measured by its transpose, we obtain a $p \times p$ matrix of which a typical term is

$$
nc_{ij} = \sum_{k=1}^{n} (x_{ik} - x_{i.})(x_{jk} - x_{j.}),
\tag{1.3}
$$

where c_{ij} is the covariance of x_i and x_j. We may (measuring from the means) write this as

$$
c = \frac{1}{n}xx'
\tag{1.4}
$$

where x' denotes the transpose of x (an $n \times p$ matrix). We shall, unless the contrary is specified, use the divisor n, not $(n-1)$, in defining variance and covariance terms. c is referred to as the covariance or dispersion matrix.

In particular, if each x_i is divided by the square root of the variance of x_i, the covariances become the correlation matrix

$$
r =
\begin{bmatrix}
1 & r_{12} & r_{13} & \cdots & r_{1p} \\
r_{12} & 1 & r_{23} & \cdots & r_{2p} \\
\cdot & \cdot & \cdot & \cdots & \cdot \\
r_{1p} & r_{2p} & r_{3p} & \cdots & 1
\end{bmatrix}
\tag{1.5}
$$

The covariance matrix and the correlation matrix are symmetric about the main diagonal, i.e. $c_{ij} = c_{ji}$ and $r_{ij} = r_{ji}$. The covariance matrix and its determinant play a fundamental role in multivariate theory, analogous to that of the variance in univariate theory, and they run thematically throughout the whole subject.

1.8   The observation matrix (1.1) can be considered from two points of view, according to whether we read it horizontally or vertically. If we read it across, say by comparing two rows, we are examining the relationship between the variables in the set of n members. Read vertically, it gives us the relation between a pair of members in the set of variables.

Example 1.2

Suppose, for instance, that the observation matrix consisted of the scores of 50 students ($n = 50$) in 10 examinations ($p = 10$), English, French, History, Mathematics, Music, etc. We might be interested in two different types of question:

(a)   whether scores in one subject were associated with scores in another, e.g. do high scores in mathematics and music tend to occur together?

(b)   how the students group over the examination field, e.g. are they clearly separable into three sets, good, indifferent, bad, or should we regard them as giving a more-or-less continuous spectrum of performance, ranging from very good to very bad?

1.9   The dual way in which we can read the matrix corresponds to two completely different methods of representing the data geometrically. The first is a natural generalization of the scatter diagram which the reader has presumably met in the elementary theory of correlation and regression. We imagine p orthogonal axes. Any member of the set corresponds to a point in the p-dimensional space so determined, the co-ordinate along the ith axis being x_{ij} for the jth point. Thus the observations correspond to a swarm of points in p dimensions, and a focal interest lies in the pattern of that swarm: whether it collapses into a flat space of lower dimension; whether it is distributed more or less spherically, or alternatively in an elongated shape; whether it splits into distinct groups, and so on. For illustrative purposes we shall only be able to draw pictures in two dimensions, but our arguments will be general.

Incidentally, this geometrical way of expressing the situation does not in the least depend on transcendental considerations as to whether more than three dimensions "exist". It may be regarded as expressing in a convenient language mathematical facts which can, if necessary, be put in algebraic terms. It is easier and is more suggestive to say that a $(p - 1)$-dimensional hyperplane meets a hypersphere of p dimensions in a hypersphere of one lower dimension (a plane cuts a sphere in a circle) than to say that the common points of the locus

$$\sum_{i=1}^{p} (x_i - a)^2 = b \quad \text{and} \quad \sum_{j=1}^{p} c_j x_j = d$$

can, by a suitable co-ordinate transformation, be expressed in the form

$$\sum_{i=1}^{p-1} (y_i - k)^2 = l.$$

1.10   The second, and less familiar, geometrical representation consists of regarding the complex, not as n points in p dimensions, but as p points in n dimensions. We take n orthogonal axes, one corresponding to each member of the set, and determine p points P_i, the co-ordinates of P_i being the n values $x_{i1} - x_{i.}, \ldots, x_{in} - x_{i.}$. There is thus one point for each variable. The distance of P_i from the origin O is given by

$$OP_i^2 = \sum_{j=1}^{n} (x_{ij} - x_{i.})^2, \tag{1.6}$$

namely n times the variance of $x_{i.}$

If we standardize the variables so as to have unit variance, the p vectors OP_i all have unit length and their extremities lie on a hypersphere of unit radius. Moreover, the cosine of the angle between OP_i and OP_j, say θ_{ij}, given by

$$\cos \theta_{ij} = \sum_{k=1}^{n} (x_{ik} - x_{i.})(x_{jk} - x_{j.}) \Big/ \left\{ \sum_{k=1}^{n} (x_{ik} - x_{i.})^2 \sum_{k=1}^{n} (x_{jk} - x_{j.})^2 \right\}^{\frac{1}{2}} \tag{1.7}$$

is the correlation between x_i and x_j.

The p vectors $OP_1 \ldots OP_p$ determine a p-dimensional space embedded in the n-dimensional space with which we began. They may be pictured like the spokes of an umbrella. If two variables are highly correlated, r is near unity and θ_{ij} is near zero. If the two have zero correlation, θ_{ij} is $\frac{1}{2}\pi$ and the vectors are orthogonal. The extent to which the vectors "bunch together" is a measure of the closeness of their intercorrelation. If one vector lies in the $(p-1)$ space determined by the others, the variable concerned is a linear function of the other variables. The pattern of vectors reveals the nature of the relationship among variables in the same way that the pattern of n points in the other kind of representation reveals the relationship among individuals.

We shall refer to the representation of **1.9** as the First Kind and that of **1.10** as the Second Kind of Spatial Representation. The distinction will be familiar to psychologists, who have for many years differentiated between the so-called P-technique (correlation among variables) and the Q-technique (correlation among individuals).

1.11 One of the basic concepts of matrix theory is that of *rank*. A matrix of order $p \times n$ is said to be of rank m ($m \leqslant$ lesser of p, n) if all determinants (minors) of more than m rows and columns which can be picked out of the matrix vanish, but at least one $m \times m$ determinant does not. A standard result of matrix theory states that if a matrix is of rank m, all the values x_{ij} are linearly dependent on m sets of them, say x_{ij} for $i, j = 1, 2, \ldots, m$. In geometrical language, in the First Kind of Spatial Representation the n points ($n > p$) will lie in a space of $m \leqslant p$ dimensions.

It is also known that the rank of the product of a matrix by its transpose is equal to the rank of the matrix. In our case this is equal to the rank of the covariance or of the correlation matrix. Thus, instead of having to examine all the determinants in the original $p \times n$ matrix in order to determine rank, it is sufficient to consider the rank of the $p \times p$ covariance matrix.

Example 1.3

Consider a four-variate case with matrix

$$
\begin{array}{cccc}
1 & 0{\cdot}8 & 0{\cdot}6 & 0{\cdot}6 \\
0{\cdot}8 & 1 & 0{\cdot}96 & 0 \\
0{\cdot}6 & 0{\cdot}96 & 1 & -0{\cdot}28 \\
0{\cdot}6 & 0 & -0{\cdot}28 & 1
\end{array}
\tag{1.8}
$$

The determinant of the whole matrix will be found to vanish. So do all the 3×3 determinants obtained by striking out one row and one column. On the other hand, the 2×2 determinants do not. Thus the matrix is of rank 2 ($p = 4$, $m = 2$).

It follows that the variation, which is apparently four-dimensional, is only two-dimensional, i.e. it is possible to find new variables, say ξ_1 and ξ_2, in terms of which x_1, x_2, x_3, x_4 can be expressed. One such pair of variables (not necessarily unique) are

$$
\left.
\begin{aligned}
x_1 &= \xi_1 \\
x_2 &= 0{\cdot}8\xi_1 + 0{\cdot}6\xi_2 \\
x_3 &= 0{\cdot}6\xi_1 + 0{\cdot}8\xi_2 \\
x_4 &= 0{\cdot}6\xi_1 - 0{\cdot}8\xi_2
\end{aligned}
\right\}
\tag{1.9}
$$

We cannot express the ξ's uniquely in terms of the x's, but conversely, from (1.9),

$$
\operatorname{var} \xi_1 = \operatorname{var} x_1 = 1
$$

$$
\operatorname{var} \xi_2 = \operatorname{var}(-4/3x_1 + 5/3x_2)
$$

$$
= \tfrac{16}{9} + \tfrac{25}{9} - \tfrac{40}{9} \operatorname{cov}(x_1 x_2).
$$

$$
= \tfrac{41}{9} - \tfrac{32}{9} = 1.
$$

It is then readily verified that the correlations of the x's are in fact those of (1.8).

Example 1.4

Consider a p-dimensional complex in which all the variables are equally correlated, the correlation matrix being

$$
\begin{matrix}
1 & r & r & \ldots\ldots & r \\
r & 1 & r & \ldots\ldots & r \\
\cdot & \cdot & \cdot & \ldots\ldots & \cdot \\
r & r & r & \ldots\ldots & 1
\end{matrix}
\tag{1.10}
$$

Adding the rows, taking out a factor in $\{1 + (p-1)r\}$ and then subtracting r times the unit row from all other rows, we find for the determinant of this matrix

$$
(1-r)^{p-1}\{1 + (p-1)r\}.
\tag{1.11}
$$

This cannot vanish unless $r = 1$ (a trivial case) or $r = -1/(p-1)$. Except in these special cases the rank of the matrix must be p, and hence we cannot represent a set of equally correlated variables in fewer than p dimensions.

1.12   It is of some interest to record a result concerning the rank of a *symmetric* matrix which relieves us of the necessity for testing every minor: if one m-rowed principal minor (i.e. a minor whose diagonal elements are taken from the diagonal elements of the original matrix) is not zero and that minor vanishes (a) when any row and the corresponding column is annexed to it and (b) when any two rows and the corresponding columns are annexed to it, then the rank is m.

There are p rows and a new row can be annexed to m in $p-m$ ways, and two rows in $\frac{1}{2}(p-m)(p-m-1)$ ways. The number of conditions on a symmetric matrix for it to be of rank m is then $\frac{1}{2}(p-m)(p-m+1)$. These are, in fact, independent conditions (cf. Lederman, 1937).

1.13   In arithmetical practice it is exceptional for the covariance determinant to vanish, but on many occasions it may be quite small. As we shall see, a number of multivariate operations depend on the inversion of the covariance matrix, that is to say, on the reciprocal of its determinant. Such operations are endangered when the determinant is small, especially for determinants of high dimensions where repetitive procedures can furnish cumulative errors. We shall revert to this topic later in the context of regression analysis (Chapter 7). It is an important instance of the point made in 1.5 (2) about the reliability of machine programs.

NOTES

(1)   A question very frequently asked, but very infrequently **answered, is:** what should be the arithmetic relation, if any, between the number of observations n and the dimension number p? On the face of it, one would require n to be considerably greater than p — if it is less, there must be collinearities in the covariance matrix and some dimensions are nugatory. But even here, cases arise where the statistician has to try to extract some sort of sense from the situation $n < p$, as for example when a large number of observations such as symptoms are recorded on relatively few patients. It is not even obvious that the number p should be proportional to n — we noticed in the text that the number of covariance terms is of the order of $\frac{1}{2}p^2$. As a rough working rule I like to have at least ten times as many observations as dimensions, but this is a most subjective personal opinion.

(2)   Although two-dimensional projections of p-dimensional complexes can be misleading, it is usually worth while to draw some of the $\frac{1}{2}p(p-1)$ possible scatter diagrams, not so much to study the condensation of a swarm as to look for outlying observations, which may not be genuine and which in any case exert an undue influence on the values of correlations or regressions.

(3)   An ingenious attempt to represent points in p-dimensional space by curves in two dimensions is made by Andrews (1972). Given a point in the First Space with co-ordinates $x_1, \ldots, x_p$, construct

$$f_x(t) = x_1/\sqrt{2} + x_2 \sin t + x_3 \cos t + x_4 \sin 2t + x_5 \cos 2t + \ldots$$

This function, plotted over the range $-\pi$ to $+\pi$, corresponds to the point. If two points are close together, so will be the corresponding curves. The method, so to speak, explores the space with a roving vector whose direction is a function of t.

(4)   The diagram overleaf may help to exhibit the coverage of the subject and the relationships among its component parts.

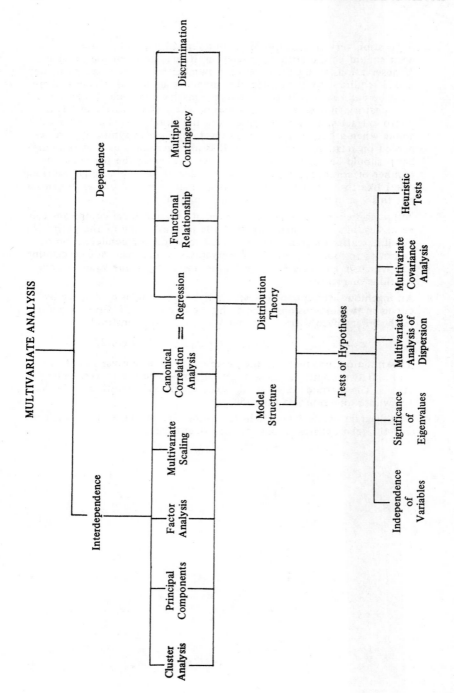

2

Principal components

2.1   Suppose that we have observations of p variables x on each of n individuals. We shall, as a first attempt to simplify the situation, inquire whether it is possible to find some new variables $\xi_1, \xi_2, \ldots, \xi_p$ which are linear functions of the x's but are themselves uncorrelated. In fact we look for p^2 constants l_{ij} $(i, j = 1, \ldots, p)$ such that

$$\xi_i = \sum_{j=1}^{p} l_{ij} x_j. \tag{2.1}$$

If the x's are measured from their means, the sum of any x_j over the n individuals is zero and hence, from (2.1), the same is true of any ξ_i. We now impose the condition that the ξ's be uncorrelated, i.e. that

$$E\left(\xi_i \xi_j\right) = E\left\{\sum_{k=1}^{p} l_{ik} x_k \sum_{m=1}^{p} l_{jm} x_m\right\} = 0, \quad i \neq j. \tag{2.2}$$

This is equivalent to

$$\sum_{k,m=1}^{p} l_{ik} l_{jm} E(x_k x_m) = \sum_{k,m=1}^{p} l_{ik} l_{jm} c_{km} = 0, \tag{2.3}$$

where c_{km} is the covariance of x_k and x_m. We note from (2.2) that this imposes $\frac{1}{2}p(p-1)$ conditions on the l's. There are p^2 of them, so we can impose a further $p^2 - \frac{1}{2}p(p-1)$ conditions and still find a solution. We shall, in fact, impose the condition that the transformation is self-orthogonal, i.e. that

$$\sum_{i=1}^{p} l_{ij} l_{ik} = 0, \quad j \neq k$$
$$= 1, \quad j = k. \tag{2.4}$$

This just absorbs the $\frac{1}{2}p(p+1)$ degrees of freedom in the transformation, so that under these conditions there will, in general, be only a finite set of solutions.

2.2   Geometrically, the orthogonality conditions (2.4) mean that in the First Space we are looking for a rotation of the co-ordinate axes about the origin. The relations are most conveniently expressed in matrix form. Representing the x's by the column ($p \times 1$) vector x and similarly for the ξ's, we have, corresponding to (2.1),

$$\xi = lx. \tag{2.5}$$

The orthogonality conditions (2.4) are equivalent to

$$ll' = I, \tag{2.6}$$

where I is the identity matrix (i.e. a $p \times p$ matrix with units in the diagonals and zeros elsewhere). If we premultiply (2.6) by l' and postmultiply by the inverse of l', we have equivalently

$$l'l = I. \tag{2.7}$$

It will follow from (2.5) that

$$x = l'\xi. \tag{2.8}$$

For the covariances of the ξ's we have, corresponding to (2.3),

$$E\,(\xi\xi') \;=\; E\,\{lx(lx)'\} \;=\; E\,(lxx'l') \;=\; lcl' \;=\; \Lambda, \quad \text{say,} \tag{2.9}$$

where Λ is a matrix with zero off-diagonal elements and with diagonal elements $\lambda_1, \lambda_2, \ldots, \lambda_p$ representing the variances of the ξ's. Premultiplying (2.9) by l', we find that

$$cl' = l'\Lambda. \tag{2.10}$$

In full this is

$$
\begin{bmatrix}
c_{11} & c_{12} & \cdots & c_{1p} \\
c_{21} & c_{22} & \cdots & c_{2p} \\
\cdot & \cdot & \cdots & \cdot \\
c_{p1} & c_{p2} & \cdots & c_{pp}
\end{bmatrix}
\begin{bmatrix}
l_{11} & l_{21} & \cdots & l_{p1} \\
l_{12} & l_{22} & \cdots & l_{p2} \\
\cdot & \cdot & \cdots & \cdot \\
l_{1p} & l_{2p} & \cdots & l_{pp}
\end{bmatrix}
$$

$$
=
\begin{bmatrix}
l_{11} & l_{21} & \cdots & l_{p1} \\
l_{12} & l_{22} & \cdots & l_{p2} \\
\cdot & \cdot & \cdots & \cdot \\
l_{1p} & l_{2p} & \cdots & l_{pp}
\end{bmatrix}
\begin{bmatrix}
\lambda_1 & & & \\
& \lambda_2 & & \\
& & \cdot & \\
& & & \lambda_p
\end{bmatrix}
\tag{2.11}
$$

Expanded in full this would give us p^2 equations. Consider the subset of p derived from the first column in the matrix product:

$$c_{11}l_{11} + c_{12}l_{12} + \ldots + c_{1p}l_{1p} = l_{11}\lambda_1$$
$$c_{21}l_{11} + c_{22}l_{12} + \ldots + c_{2p}l_{1p} = l_{12}\lambda_1$$
$$\cdot \quad \cdot \quad \cdot \cdots \quad \cdot \quad \cdot$$
$$c_{p1}l_{11} + c_{p2}l_{12} + \ldots + c_{pp}l_{1p} = l_{1p}\lambda_1 \tag{2.12}$$

or

$$(c_{11} - \lambda_1)l_{11} + c_{12}l_{12} + \ldots + c_{1p}l_{1p} = 0$$
$$c_{21}l_{11} + (c_{22} - \lambda_1)l_{12} + \ldots + c_{2p}l_{1p} = 0$$
$$\cdot \qquad \cdot \qquad \cdot \qquad \cdots \qquad \cdot$$
$$c_{p1}l_{11} + c_{p2}l_{12} + \ldots + (c_{pp} - \lambda_1)l_{1p} = 0. \qquad (2.13)$$

Eliminating the l's, we have

$$\begin{vmatrix} c_{11} - \lambda_1 & c_{12} & \cdots & c_{1p} \\ c_{21} & c_{22} - \lambda_1 & \cdots & c_{2p} \\ \cdot & \cdot & \cdots & \cdot \\ c_{p1} & c_{p2} & \cdots & c_{pp} - \lambda_1 \end{vmatrix} = 0, \qquad (2.14)$$

or, in matrix notation,

$$|c - \lambda_1 I| = 0. \qquad (2.15)$$

It will be found that the other equations in $\lambda_2, \ldots, \lambda_p$ are of exactly the same kind. Thus the required λ's (the variances of the ξ's) are the p roots of

$$|c - \lambda I| = 0. \qquad (2.16)$$

The λ's are, in fact the so-called *eigenvalues* of the matrix c. The corresponding l's are the components of the *eigenvector*.

2.3   To find the required transformation we then solve (2.16) for the values of λ. Then, going back to equations (2.13), together with the condition that $\sum_{i=1}^{p} l_{ij}^2 = 1$, we can solve for the l's and hence determine the transformation. It is, except in degenerate cases which we shall notice in a moment, unique except for the signs of the l's (which merely correspond to a reversal of the direction of the corresponding eigenvector).

In pre-computer times the arithmetic involved in the solution of the equations was severe. Nowadays most sizeable computers have associated with them a machine program which performs it very quickly, and we need not pause to discuss the solutional procedure, except to remark that some programs are better than others.

2.4   The new functions ξ are known as *principal components*. They are uncorrelated linear functions of the original variables. In particular, if the original variables were normally distributed, the new variables ξ are not only uncorrelated but independent. It is remarkable that they have another useful property. Suppose, in fact, that we had approached the subject from a different viewpoint and looked for the linear functions of the x's which have stationary values of their variance. If one of these is ζ, expressed as

$$\zeta = \sum_{j=1}^{p} l_j x_j,$$
(2.17)

its variance is given by

$$\text{var } \zeta = \sum_{j=1}^{p} \sum_{k=1}^{p} l_j l_k \text{ cov } (x_j x_k).$$
(2.18)

We require to find the stationary value of this, subject to the condition

$$\sum_{j=1}^{p} l_j^2 = 1.$$
(2.19)

This is equivalent to an unconditional stationary value of

$$\Sigma \, l_j l_k c_{jk} - \lambda(\Sigma \, l_j^2 - 1),$$
(2.20)

where λ is an undetermined multiplier. Differentiation by l_j leads to

$$\sum_{k=1}^{p} l_k c_{jk} - \lambda l_j = 0.$$
(2.21)

This leads to equation (2.16) with λ as one of the eigenvalues. In other words, the linear combinations with stationary values are the principal components. If we christen the components according to the magnitude of their eigenvalues (variances), we may say that the first component, ξ_1, has the largest possible variance of any linear function; the second, ξ_2, has the largest variance subject to being uncorrelated with the first; the third has the largest variance of linear functions which are uncorrelated with the first and second; and so on down to the smallest eigenvalue, which corresponds to the linear combination with the smallest variance.

2.5   It is interesting to consider the same problem from the geometrical point of view. Consider the n points in the First Space and the line with current co-ordinates X

$$\frac{X_1 - g_1}{l_1} = \frac{X_2 - g_2}{l_2} = \ldots = \frac{X_p - g_p}{l_p},$$
(2.22)

where the l's are direction cosines and are therefore subject to the condition

$$\sum_{i=1}^{p} l_i^2 = 1.$$
(2.23)

The projection of a vector from the point $(g_1, g_2, \ldots, g_p)$ to the jth point on to this line is

$$\sum_{i=1}^{p} l_i (x_{ij} - g_i).$$

Thus the sum of the squares of distances of all the points from the line, say nS, is given by

$$nS = \sum_{j=1}^{n} \left[\sum_{i=1}^{p} (x_{ij} - g_i)^2 - \left\{ \sum_{i=1}^{p} l_i(x_{ij} - g_i) \right\}^2 \right]. \qquad (2.24)$$

If this is a stationary value the partial differentials with respect to g_i vanish, leading to

$$\sum_{j=1}^{n} (x_{ij} - g_i) - \sum_{j=1}^{n} l_i \sum_{i=1}^{p} l_i(x_{ij} - g_i) = 0. \qquad (2.25)$$

If we choose our origin at the mean of the x's, the terms in x vanish under summation with respect to j, and we find that

$$-ng_i + l_i n \sum_i l_i g_i = 0$$

or
$$\frac{g_i}{l_i} = \text{constant}. \qquad (2.26)$$

Hence, from (2.22), the origin lies on the line, and without loss of generality we may take the g's equal to zero. This is equivalent to saying that the line we are seeking goes through the centre of gravity of the points.

Equation (2.24) then reduces to

$$nS = \sum_{j=1}^{n} \left\{ \sum_{i=1}^{p} x_{ij}^2 - \left(\sum_{i=1}^{p} l_i x_{ij} \right)^2 \right\} \qquad (2.27)$$

$$= n \sum_{i=1}^{p} c_{ii}^2 - n \sum_{i,k=1}^{p} l_i l_k c_{ik}. \qquad (2.28)$$

If we find the stationary values of this, subject to (2.23), we are led once again to equations (2.21) and (2.16), exactly as in 2.4. In particular, from (2.21) multiplied by l_j and summed, we have

$$\lambda = \sum_{j,k=1}^{p} l_j l_k c_{jk},$$

so that, from (2.28), the largest λ corresponds to the least S. In other words the line corresponding to the largest eigenvector is the one for which the sum of squares of distances from the n points on to it is a minimum. Furthermore, if we project the points on to a hyperplane perpendicular to the line, the second eigenvalue corresponds to a line in that hyperplane such that the sum of squares of distances on to it is a minimum; and so on.

2.6 The determination of the first principal compound may therefore be regarded as finding the line of closest fit to the n points in the First Space. Hence "closest" is used in the sense of minimal sum of squares of distances from the line. It is to be sharply distinguished from regression lines, for which we minimize sums of squares in particular directions.

2.7 It is also of interest to observe that in the Second Space the vectors

the principal components are themselves orthogonal, as is obvious
that we have constructed them to be uncorrelated. Thus the
~ing the components in the First Space, and the vectors rep-
~~~~~~ung them in the Second Space, both form an orthogonal set.

2.8   We remark on a few subsidiary points without detailed proof.

(1)   The matrix c is special in the sense that it is symmetric and, being a
covariance matrix, is non-negative definite. This implies that the eigenvalues are always real and never negative. If the practitioner arrives at
cases where negative or imaginary eigenvalues appear, he should look
for the method used to generate the covariances, especially if the matrix
is standardized to unit variance and the covariances are correlations.
Cases occur in which correlations are not calculated by ordinary product-
moment methods but are "estimated" from 2×2 association tables by
other means. This may give rise to impossible eigenvalues.

(2)   In degenerate cases some of the λ's may be equal. Indeed, in the extreme case they may all be equal. There is then some degree of indeterminacy in the eigenvectors and the transformation is not unique. For
example, if the original x's are uncorrelated, the covariance matrix becomes diagonal and the λ's are simply the variances of the x's, with
which the ξ's coincide; no transformation is required. If in addition
the x's have been standardized to unit variance, any orthogonal transformation to ξ's will result in another set of uncorrelated variables with
unit variance.

(3)   The sum of squares of the distances of all the points from their centre
of gravity may be expressed as the sum of the eigenvalues. If this sum
is S, we may, in a sense, say that the first component "accounts for"
a proportion λ_1/S of the total variation, the first two for proportion
$(\lambda_1 + \lambda_2)/S$, and so on. This is not an analysis of variance in the ordinary sense of the term, but it is convenient to use expressions such as
"the first four components account for P percent of the variation".

2.9   In particular, if certain λ's vanish, say $p - m$ of them, the correlation
matrix is of rank m and the variation collapses into a space of $m < p$ dimensions. In arithmetical practice it is rare for the lower eigenvalues to vanish
exactly, but not uncommon for some of them to be small. In such a case
we may say that the variation falls *approximately* into a space of lower dimension. Geometrically one may picture the situation as one in which a
p-dimensional complex has slight variation in one or more directions, as
though a globular cluster had flattened out into a pancake-shaped distribution,
and we can ignore the variation perpendicular to the plane of the pancake.
In some circumstances we may then discard those dimensions in which the
variation is small.

This is not the same thing as discarding the variables themselves. The eigenvectors corresponding to small λ's may nevertheless be functions of all the original variables with non-negligible coefficients. We must defer a discussion of discarding procedures until a later chapter.

2.10   There is one distinctive feature of component analysis which seriously affects the use which we make of it in practice. The components are not independent of the scales in which we measure the original variables. Consider, for example, the simple case of two variables x_1 and x_2. If we were regressing one on the other it would be a matter of indifference what scales we used; or, to put it another way, the correlations are invariant under changes of scale. This is not true of the principal components. We shall illustrate the point numerically below. Algebraically it is evident that if we alter the scale of some of the variables in (2.16) we get a different determinantal equation and therefore different values of the λ's. Geometrically, from the viewpoint of **2.5** the shortest distance from a point to a line is not invariant under the changes of scale of some of the variables (right angles do not project into right angles).

2.11   This difficulty is often overcome by standardizing the variables from the outset to have unit variance. The characteristic equation (2.16) then becomes

$$|\mathbf{r} - \lambda \mathbf{I}| = 0. \tag{2.29}$$

Cases occur, of course, in which all the variables are in comparable scales from the outset: e.g. in psychology, where the primary data may be scores on a range from 0 to 100; or in meteorology, where all the measurements may be in degrees centigrade; or in anthropometry, where all the variables may be in millimetres. But even here we have to be a little careful in working with unstandardized data. For example, in anthropometry we may have measurements on the length of the foot, say from 15 to 30 centimetres, and on the overall height, say from 150 to 200 centimetres; in such a case it clearly makes a good deal of difference whether we calculate from the data pre-standardized to unit variance or from the original measurements.

Example 2.1

Consider again the situation of Example 1.4, in which all the variables are equally correlated. The equation for the eigenvalues is now

$$\begin{vmatrix} 1-\lambda & r & r & \ldots & r \\ r & 1-\lambda & r & \ldots & r \\ \cdot & \cdot & \cdot & \ldots & \cdot \\ r & r & r & \ldots & 1-\lambda \end{vmatrix} = 0. \tag{2.30}$$

In the manner of Example 1.4 we reduce this to

$$\{1 - \lambda + (p - 1)r\}\{1 - \lambda - r\}^{p-1} = 0.$$

Thus we have one eigenvalue equal to $1 + (p - 1)r$, and all the others equal to $1 - r$. We note in passing that the sum of the eigenvalues is equal to p, the trace (sum of the diagonal elements) of the covariance matrix. The variation may be pictured as cigar-shaped, with one long axis corresponding to the major eigenvalue, and the variation orthogonal to that axis being spherically homogeneous.

Example 2.2

Table 2.1 gives some data relating to 20 samples of soil, i.e. sand content (x_1), silt content (x_2), clay content (x_3), organic matter (x_4), and acidity on the pH scale (x_5). The data are in fact a subset of a much larger batch of material but have been confined to 20 to economize on arithmetic.

Table 2.1   *Samples of soil*

Soil number	x_1 Sand content %	x_2 Silt content %	x_3 Clay content %	x_4 Organic matter %	x_5 pH
1	77·3	13·0	9·7	1·5	6·4
2	82·5	10·0	7·5	1·5	6·5
3	66·9	20·6	12·5	2·3	7·0
4	47·2	33·8	19·0	2·8	5·8
5	65·3	20·5	14·2	1·9	6·9
6	83·3	10·0	6·7	2·2	7·0
7	81·6	12·7	5·7	2·9	6·7
8	47·8	36·5	15·7	2·3	7·2
9	48·6	37·1	14·3	2·1	7·2
10	61·6	25·5	12·9	1·9	7·3
11	58·6	26·5	14·9	2·4	6·7
12	69·3	22·3	8·4	4·0	7·0
13	61·8	30·8	7·4	2·7	6·4
14	67·7	25·3	7·0	4·8	7·3
15	57·2	31·2	11·6	2·4	6·5
16	67·2	22·7	10·1	3·3	6·2
17	59·2	31·2	9·6	2·4	6·0
18	80·2	13·2	6·6	2·0	5·8
19	82·2	11·1	6·7	2·2	7·2
20	69·7	20·7	9·6	3·1	5·9

We note that the variation of x_1 is a good deal larger than that of x_4, so we may expect scaling problems. Let us in the first instance perform a principal component analysis on the unstandardized data. The covariance matrix is given in Table 2.2 and the eigenvalues and eigenvectors in Table 2.3.

Table 2.2   *Covariance matrix of data of Table 2.1*

	x_1	x_2	x_3	x_4	x_5
x_1	138·3267	− 102·1227	− 36·2040	− 0·9422	− 0·1358
x_2		79·7382	22·3846	1·5266	0·1108
x_3			13·8194	− 0·5844	0·0250
x_4				0·6434	0·0334
x_5					0·2626

Table 2.3   *Eigenvalues and eigenvectors of covariance matrix of Table 2.2*

	Eigenvalues	Coefficients				
		x_1	x_2	x_3	x_4	x_5
1	223·84	0·785	− 0·587	− 0·198	− 0·007	0·001
2	8·22	− 0·223	− 0·561	0·784	− 0·146	− 0·002
3	0·47	− 0·027	− 0·086	0·113	0·980	0·137
4	0·26	− 0·004	0·001	− 0·014	0·136	0·991
5	0·00	0·577	0·577	0·577	0·000	0·000
	232·79					

A glance at the eigenvalues reveals that one is zero. There must, therefore, be a linear relation among the variables. From the coefficients we see that the relation is

$$0{\cdot}577(x_1 + x_2 + x_3) = 0. \tag{2.31}$$

Measured about their means, the sum of the three variables is zero. We then see, on referring back to the original data of Table 2.1, that in fact the three variables add up to 100 percent. It appears that sand and clay content were measured and silt was merely obtained by subtracting their sum from 100 (or at any rate, one of them determined as the complement of the other two). We can then discard one variable and we shall, on arbitrary grounds, discard x_1.

We repeat the analysis on x_2 to x_5, obtaining Table 2.4. No eigenvalue vanishes, although two are relatively small. The position is dominated by the first eigenvalue, which accounts for over 90 percent of the variation. The eigenvector is given by

$$\xi_1 = 0.956x_2 + 0.294x_3 + 0.015x_4 + 0.001x_5. \qquad (2.32)$$

The question now arises: can we attribute any meaning to this component, or is it just an artefact conjured up by the analysis? This is the most difficult question to answer in component analysis, and one to which there is

Table 2.4   *Eigenvalues, omitting variable 1, of the covariance matrix of the data of Table 2.1*

	Eigenvalues	Coefficients			
		x_2	x_3	x_4	x_5
1	86·640	0·956	0·294	0·015	0·001
2	7·094	−0·288	+0·945	−0·154	−0·002
3	0·471	−0·059	0·142	0·979	0·137
4	0·258	0·006	−0·018	−0·136	0·991
	94·463				

rarely a simple unambiguous reply to be given. From the size of the coefficients of x_4 and x_5 it looks as if this component, if it has any "reality", is simply the physical quality of the soil, expressible in terms of silt and clay (or sand) and independent of organic matter and acidity. We might provisionally call it silt content. The second component accounts for very much less, only about 7·5 percent of the variation, and the coefficients would lead us to regard it as dominated by x_3, namely clay.

Now suppose that we work with the correlation matrix instead of the covariance matrix. The former is given in Table 2.5 and the component analysis of the variables x_2 to x_5 is displayed in Table 2.6. We now observe that

Table 2.5   *Correlation matrix of the data of Table 2.1*

	x_1	x_2	x_3	x_4	x_5
x_1	1·0000	−0·9724	−0·8281	−0·0999	−0·0225
x_2		1·0000	0·6743	0·2131	0·0242
x_3			1·0000	−0·1960	0·0131
x_4				1·0000	0·0787
x_5					1·0000

the first component accounts for 42 percent of the variation, the second for 29 percent and the third for 24 percent. This is quite a different picture from the previous one. The components are different in magnitude and the coefficients of the eigenvectors are different from those previously obtained, sometimes different even with respect to sign.

Table 2.6 *Eigenvalues, omitting variable 1, of the correlation matrix of the data of Table 2.1*

	Eigenvalues	Coefficients			
		x_2	x_3	x_4	x_5
1	1·676	0·710	0·702	0·025	0·042
2	1·146	0·182	−0·241	0·836	0·459
3	0·960	−0·147	0·111	−0·423	0·887
4	0·218	−0·664	0·661	0·349	−0·026
	4·000				

2.12 Not all situations are as sensitive to scale-changes as the one we have just considered, but the problem of scale is obviously a severe one if we are trying to attribute any kind of reality to the components. This question as to the "meaning" of components is one which we shall have to consider again in connection with factor analysis. It arises even when scaling is not a difficulty. To lay down any broad lines regarding inference from component analysis regardless of the circumstances of the individual case is impossible. The following examples will illustrate some of the uses which can fruitfully be made of it.

Example 2.3 (Stone, 1947)

Stone took Kuznets' and Barger's data for the USA comprising, for each of the years 1922-1938, 17 series regarded as constituent elements of total national income or outlay, e.g. employers' compensation, consumers' perishable goods plus producers' durable goods, net public outlay, net increase in inventories, dividends, interest, foreign balance, and so on. All figures were in dollars, but the standard deviations of the variables varied considerably, so Stone worked on the correlation matrix, which is set out in full in his paper. He then did a principal component analysis on the observations taken about their mean and extracted three principal components with $\lambda = 80·76$, 10·59, 6·09 percent, accounting for 97·44 percent of the variance. Evidently these three components account for nearly all the variation, and we have thus reduced the effective dimensions of variation from 17 to 3. This illustrates the economy in effective dimension number which it is the object of component analysis to achieve. We must remember, however, that there were only 17 observations and that they are highly correlated from year to year. Stone had reason to suppose on economic grounds that the variation was mostly accounted for by three components: (a) total income i or some similar quantity, (b) rate of change of i, say Δi, and (c) a trend term expressing expansion or contraction of the economy, which we may take as a linear term in the time t. Moreover, these quantities were separately measurable.

Stone correlated his three principal components, say F_1, F_2, F_3 (in his notation), with i, Δi and t to obtain

	F_1	F_2	F_3	i	Δi	t
F_1	1					
F_2	0	1				
F_3	0	0	1			
i	0·995	−0·041	0·057	1		
Δi	−0·056	0·948	−0·124	−0·102	1	
t	−0·369	−0·282	−0·836	−0·414	−0·112	1

The three underlined figures stand out, and it seems very reasonable to identify F_1 with i, F_2 with Δi and F_3 with t.

It is to be noticed that in this example the three variables which we are able to identify with the three principal components were independently measurable. We are therefore on much firmer ground than if they were themselves variables used in the principal component analysis.

Example 2.4 (Kendall, 1939)

The yields of ten crops were recorded for 48 counties in England ($n = 48$, $p = 10$). The crops were wheat, barley, oats, beans, peas, potatoes, turnips, mangolds, hay (temporary grass) and hay (permanent grass). The correlations between the various crops were nearly all positive, suggesting that there might be some quality "productivity" associated with an area, irrespective of the crops actually grown. To allow for climatic variations four years were chosen; the results for them agreed quite closely.

The correlation matrix was computed and the largest root ascertained. For 1925, for instance, $\lambda = 4·760$ and thus the corresponding ξ accounts for 47·6 percent of the total variation. The corresponding vector was given by

$$\xi_1 = 0·39x_1 + 0·37x_2 + 0·39x_3 + 0·27x_4 + 0·22x_5 + 0·30x_6$$
$$+ 0·32x_7 + 0·26x_8 + 0·24x_9 + 0·34x_{10}. \tag{2.33}$$

This variable was provisionally identified with productivity, and the counties were arranged in order according to the magnitude of ξ as given by (2.33). They were then grouped into very good, good, moderate, poor, bad, according to the values of ξ which they bore. The results agreed with general knowledge about the geographical distribution of productivity, except in one or two instances. Here also there were extraneous data supporting the conclusion; for example, the total cash value of the crops for each county and the calorie equivalent.

In this case a value of $\lambda = 4\cdot76$ is not very high and we suspect that another variable, at least, is required to "explain" the variation. A more detailed analysis is given by Banks, C.H. (1954). We note that all coefficients in (2.33) are positive.

From one point of view (2.33) may be regarded as determining an index ξ of productivity. In the ordinary way one might expect the crop yields to be weighted in some manner according to the production or acreage of the various crops.

2.13 Let us pursue this last point a little further. In the behavioural sciences, especially in economics, we are frequently compelled to summarize a highly complex numerical aggregate in the form of an index-number, forcing a p-dimensional system, so to speak, into one dimension. Familiar examples are index-numbers of prices, money wage-rates, cost of living, business activity, and so forth. Such index-numbers are usually constructed by weighting the constituent items by quantities which, in some sense, reflect their relative importance. We may, however, approach the subject from the point of view of principal components and ask: if the variation is to be summarized as nearly as possible in a linear combination of the variables, which is the best linear function? From this angle the first principal component, which is an answer to the question, furnishes its own weights. No current index, so far as I know, uses the principal component method, although Rhodes (1937) pointed out the possibility a long time ago. The reasons are not far to seek. The usual type of index, based on fixed weights (such as the "basket-of-goods" type familiar in cost-of-living inquiries) is much more easily understood and is more easily calculated than the sophisticated principal component. Moreover, some of the coefficients in a principal component may be negative, which might indicate to the layman that the corresponding variables are deleterious and should be omitted. However, now that the computer has removed most of the arithmetical terrors, it would in many cases be interesting to calculate the first principal component and its variance and to compare with the variance of the index in actual use. This at least would give some measure of the extent to which the actual index falls short of optimality.

2.14 If the first principal component is acceptable as a summary of the complex under study, it can be used to arrange the n members in order according to the values which they bear of the first variable ξ_1. It is of interest to note that this ordering can often be carried out by ranking methods without the calculation of eigenvalues. In the crop-productivity inquiry of Example 2.4, for example, the counties were ranked from 1 to 48 for each of the 10 crops, and the 10 ranks summed for each county. The resulting numbers gave an order which was almost identical with that provided by the first principal component. This may be regarded as a distribution-free type of component, but the method does not appear to lend itself to the detection of further components. On the other hand, it may be preferable in the determination of an ordering in those cases where the variables have a wide non-normal scatter, as for example if we are arranging the countries of the world according to an index of business activity.

The summation of ranks maximizes the average Spearman correlation between the ranking so reached and the rankings according to the p variables (Kendall, 1970). The rank-vector therefore gets as close to the p rank-vectors as it can, so to speak, and will approximate to the rank-vector given by the first principal component.

Example 2.5

Stamp (1952) applied the ranking method to an agricultural grading of certain countries. The average ranks on nine crops (wheat, rye, barley, oats, corn, potatoes, sugar beet, beans and peas) were as follows:

1934-8		1946	
Belgium	2·2	Belgium	2·3
Denmark	2·6	Denmark	2·4
Netherlands	2·9	Netherlands	2·4
Germany	4·3	New Zealand	4·2
Britain	4·7	Britain	4·8
Ireland	4·7	Ireland	5·3
New Zealand	5·8	Egypt	6·2
Egypt	6·3	Germany	7·6
Austria	7·2	USA	8·2
France	9·2	France	9·0
Japan	10·4	Canada	9·1
Italy	12·0	Austria	11·2
USA	12·0	Chile	11·5
Canada	12·3	Argentina	12·4
Spain	12·6	China	12·7
Chile	12·9	Italy	12·7
China	13·6	Japan	14·1
Argentina	14·3	Spain	14·2
Australia	16·0	India	17·0
India	17·8	Australia	17·2

With all its imperfections, this is an interesting table. The alteration in relative positions before and after World War II are like movements in a league table and form the basis of the same type of analytical commentary.

2.15 We must now pay some attention to sampling problems: if our n members are a random subset of a larger population, how reliable are the components derived from the sample in reflecting the properties of the population?

If we work from the correlation matrix of order p, the sum of the eigenvalues, as we have remarked, is p, so that the average eigenvalue is unity. It has therefore become the practice to retain as "important" those components with $\lambda > 1$ and to neglect the others. This is a very rough-and-ready procedure for which it is difficult to advance a convincing theoretical justification. If the significance of components is to be judged on a subjective basis it is better to look at the pattern of them all (which, *inter alia*, implies that one should not use programs which print out only the eigenvectors with $\lambda > 1$ and suppress the rest).

2.16 The sampling theory of eigenvalues and eigenvectors is very complicated and we must defer a discussion until Chapter 6. At this point, however, we may derive some simple large-sample results which are useful in practice in guiding our decisions whether to ignore components or not.

Since the l's are othogonal, we have

$$\sum_{j=1}^{p} l_{ji} l_{jk} = 0, \quad i \neq k$$
$$= 1, \quad i = k, \tag{2.34}$$

which we can summarize as

$$\sum_{j=1}^{p} l_{ji} l_{jk} = \delta_{ik}, \tag{2.35}$$

where δ_{ik} is a symbol (the Kronecker delta) which is unity if $i = k$ and zero otherwise. Likewise from (2.21) we derive for the ith eigenvalue

$$\sum_{j} l_{ij} c_{jk} = \lambda_i l_{ik} \tag{2.36}$$

$$\sum_{k} c_{ij} l_{jk} l_{km} = \lambda_i \delta_{jm}. \tag{2.37}$$

Taking differentials (small variations) of (2.36), we have

$$\sum_{j} c_{jk} dl_{ij} + \sum_{j} l_{ij} dc_{jk} = \lambda_i dl_{ik} + l_{ik} d\lambda_i. \tag{2.38}$$

Without loss of generality we may now suppose the variables transformed to the ξ axes, in which case $c_{jj} = \lambda_j$ and $c_{ij} = 0, i \neq j$. Then the first terms on each side of (2.38) cancel when $k = i$ and we have

$$dc_{ii} = d\lambda_i. \tag{2.39}$$

Thus
$$\text{cov}\,(\lambda_i, \lambda_j) = \text{cov}\,(c_{ii}, c_{jj}).$$

We shall prove later that cov (c_{ii}, c_{jj}) is $2c_{ij}^2/n$. Hence

$$\text{cov}(\lambda_i, \lambda_j) = 0, \quad i \neq j \tag{2.40}$$

$$\text{var } \lambda_i = \frac{2\lambda_i^2}{n}. \tag{2.41}$$

The eigenvalues are uncorrelated for large samples. The standard error of an eigenvalue is $\lambda_i(2/n)^{\frac{1}{2}}$.

2.17 Formulae can also be derived for the eigenvalues of a *correlation* matrix and for the individual l's. They are, however, very complicated and hardly ever required. (They are given in Kendall and Stuart, vol. 3, 2nd edn, p. 294.) It will be clear from the nature of maximization itself that the individual l's are rather unreliable. Consider, for example, the fitting of the first principal component in the First Space by minimizing the sum of squares of distances from it. Evidently the line can pivot on its centre of gravity to some extent without lowering the sum of squares very much (the rate of change at the minimal position, by definition, being zero). The direction cosines of the lines are therefore rather sensitive to fluctuations such as one would get from one sample to another. This is especially true where two eigenvalues are close together. It would, in my view, be unwise to lean heavily on the numerical value of any particular coefficient in the eigenvector.

2.18 The subject as we have developed it is concerned with components which are linear in the original variables. The question arises as to what should be done if the data are obviously not linear. The golden rule in such cases is to try to linearize the data from the outset, e.g. by transformations of the square-root or logarithmic type or more complicated functions if necessary. One can usually find a transformation which will linearize one pair of variables; it is not so easy to find one which will linearize them all. A certain amount of trial and error is often required in order to determine appropriate transformations.

Consider, for example, the complications which arise if we look for new variables which are quadratic functions of the old. Take

$$\xi_1 = l_1 x_1 + l_2 x_2 + \ldots + l_p x_p + m_{11} x_1^2 + m_{12} x_1 x_2 + \ldots + m_{pp} x_p^2. \tag{2.42}$$

Writing $m_{00} = 0$, x_0 as a dummy variable, and l_j as m_{0j}, we may write this as

$$\xi_1 = \sum_{j=0}^{p} \sum_{k=0}^{p} m_{jk} x_j x_k. \tag{2.43}$$

Then the mean of ξ_1 is given by

$$E(\xi_1) = \sum m_{jk} c_{jk}$$

and the mean square by

$$E(\xi_1^2) = \sum_{j,k,l,r} m_{jk} m_{lr} c_{jklr}, \tag{2.44}$$

where c_{jklr} is a fourth-order moment of the x's. Thus

$$\text{var } \xi_1 = \sum_{j,k,l,r} m_{jk} m_{lr}(c_{jklr} - c_{jk}c_{lr}). \qquad (2.45)$$

This, to be maximized subject to $\sum m_{jk}^2 = 1$, gives us

$$\sum m_{lr}(c_{jklr} - c_{jk}c_{lr}) - \lambda m_{jk} = 0. \qquad (2.46)$$

The elimination of the m's gives a determinant of $p + p + \frac{1}{2}p(p-1) = \frac{1}{2}p(p+3)$ rows and columns, dependent on quartic moments. The components, moreover, are not orthogonal. It is not surprising that this line of attack has not been pursued.

NOTES

(1)   Joliffe (1972) considers several methods of discarding *variables* in addition to the one which we shall consider in Chapter 7.

(2)   The use of principal components in regression analysis will be discussed in Chapter 7.

(3)   The ratio of the largest to the smallest eigenvalue has sometimes been used as an index of the extent to which the covariance matrix is ill-conditioned (i.e. has a determinant near to zero). I would not recommend such a measure. It takes no account of the number of small eigenvalues.

(4)   The roots of equation (2.16) are not algebraic functions of the covariances and therefore of the observations. Certain functions of the roots, however, are so. If we imagine (2.16) expanded as a polynomial in λ, say as

$$\lambda^p + a_1 \lambda^{p-1} + \ldots + a_{p-1}\lambda + a_p = 0,$$

we see that the product of the roots is a_p, obtained by putting $\lambda = 0$ in the equation. Thus the product $\lambda_1 \lambda_2 \ldots \lambda_p$ is the covariance determinant $|c|$. Further, the sum of the roots is the coefficient $-a_1$ and is seen to be the sum of the diagonal elements in c, namely the so-called "trace" of the matrix.

(5)   With reference to the comments in 2.13, an index of eutrophication (healthy nutrition) in American lakes has been constructed by principal components from certain physical variables. See Shannon, E.E. and Brezonik, P.L. (1972), *Journal of the Sanitary Engineering Division, American Society of Civil Engineers*, February, 37.

3

Classification and clustering

3.1 One of the basic problems of science in reducing the world to order (or, if you prefer it, in imposing a man-made order on the complexity of things) is to classify. Thus, given a set of n objects and observations of their qualities, we require to form them into groups on the basis of their internal similarities. The subject of classification in its broadest aspects is very wide indeed and we shall narrow the field to some extent in considering what the statistician can contribute to it. For one thing, the taxonomist often has at his disposal data of a non-statistical kind; the zoologist has records of biological lines of evolution, or genetic knowledge which influences his distinctions by genera and species; the linguist has knowledge of population movements which affect his classification of human languages. We shall not be directly concerned with these extraneous constraints on the taxonomic process, though they would have to be taken into account in the practical interpretation of our results. Ours is a simple problem: given n members on each of which are observed p characteristics, is there any evidence that they cluster into definable groups, as against the alternative hypothesis that they are an unstructured set?

3.2 There is another point of difference between classification and clustering. The basis on which we classify may vary in different sub-groups. A librarian, for example, might define a class of books on philosophy and another class of novels. But if he wished to sub-classify philosophical works his criteria would probably be different from those which he would employ in sub-classifying novels. There is always an element of arbitrariness in such a hierarchical classification, but it is an arbitrariness of convenience. In the clustering case we use the p variables throughout, although, as we shall see, some of them may be more important in delimiting certain clusters than others for different clusters.

3.3 The very concept of "cluster" is a subjective matter. In ordinary speech one thinks of a cluster as a set of objects which are "all close together" like a sunburst or a knot of people at a party. There are other systematic patterns

31

to which one might hesitate to attribute the word. The rings of Saturn, for example, are certainly a non-random grouping of particles, but it requires precision in our definitions whether to regard a toroid as a cluster. Again, the track of water droplets in a Wilson chamber, effectively a linear sequence, is certainly systematic, but a set of objects arranged along a straight line would not be described as a cluster in the ordinary connotation of the word.

3.4   The general lesson to be learned from the contemplation of such cases is that different methods may be required to detect different patterns of points. There is no one method of cluster analysis. For the most part, in the rest of this chapter, we shall be speaking of clusters in the sense of a group scattered around some central value, possibly condensing in a nuclear set, not necessarily spherical but not excessively elongated into a rod-like shape. One major difficulty, as we noticed in Chapter 1, is that for more than three dimensions it is impossible to draw a picture of the scatter of n points and determine the clusters by eye. The computer has to do this for us.

3.5   The difficulty of defining what we mean by "cluster" in terms explicit enough to be explained to a computer, which cannot cope with doubt or ambiguity, may perhaps account for the very large number of different methods that have been proposed for carrying out cluster analysis. In a useful review of the subject, Cormack (1971) listed about 200 references bearing on the problem of clustering and described some of the methods in current use. I shall not have space to describe them all. Most of them have one thought in common − the extent to which two groups are to be regarded as separate clusters depends on the distance between their means as compared to the mean distance within the clusters, a criterion bearing a formal resemblance to the test of homogeneity of classes used in the analysis of variance. One desirable feature, which not every proposed method possesses, is that of being able to handle large quantities of data, several thousands of members, or a large number of variables, say 50 or 100, with a small expenditure of computer time.

3.6   The duality between variables and members which was noticed in earlier chapters applies also in cluster analysis. We may wish to cluster variables, in the sense of examining whether a subset is so highly correlated that one of them, or an average of them, can be used to represent the set without serious loss of information; or we may wish to examine whether the individuals fall into groups. The point of making the distinction is that the clustering problem for variables is usually simpler than that for individuals, in two ways: (1) the procedure is scale-invariant, and (2) it can usually be carried out by hand on the correlation matrix.

Table 3.1 *48 applicants scaled on 15 variables*

Person	Variable														
	1	2	3	4	5	6	7	8	9	10	11	12	13	14	15
1	6	7	2	5	8	7	8	8	3	8	9	7	5	7	10
2	9	10	5	8	10	9	9	10	5	9	9	8	8	8	10
3	7	8	3	6	9	8	9	7	4	9	9	8	6	8	10
4	5	6	8	5	6	5	9	2	8	4	5	8	7	6	5
5	6	8	8	8	4	4	9	2	8	5	5	8	8	7	7
6	7	7	7	6	8	7	10	5	9	6	5	8	6	6	6
7	9	9	8	8	8	8	8	8	10	8	10	8	9	8	10
8	9	9	9	8	9	9	8	8	10	9	10	9	9	9	10
9	9	9	7	8	8	8	8	5	9	8	9	8	8	8	10
10	4	7	10	2	10	10	7	10	3	10	10	10	9	3	10
11	4	7	10	0	10	8	3	9	5	9	10	8	10	2	5
12	4	7	10	4	10	10	7	8	2	8	8	10	10	3	7
13	6	9	8	10	5	4	9	4	4	4	5	4	7	6	8
14	8	9	8	9	6	3	8	2	5	2	6	6	7	5	6
15	4	8	8	7	5	4	10	2	7	5	3	6	6	4	6
16	6	9	6	7	8	9	8	9	8	8	7	6	8	6	10
17	8	7	7	7	9	5	8	6	6	7	8	6	6	7	8
18	6	8	8	4	8	8	6	4	3	3	6	7	2	6	4
19	6	7	8	4	7	8	5	4	4	2	6	8	3	5	4
20	4	8	7	8	8	9	10	5	2	6	7	9	8	8	9
21	3	8	6	8	8	8	10	5	3	6	7	8	8	5	8
22	9	8	7	8	9	10	10	10	3	10	8	10	8	10	8
23	7	10	7	9	9	9	10	10	3	9	9	10	9	10	8
24	9	8	7	10	8	10	10	10	2	9	7	9	9	10	8
25	6	9	7	7	4	5	9	3	2	4	4	4	4	5	4
26	7	8	7	8	5	4	8	2	3	4	5	6	5	5	6
27	2	10	7	9	8	9	10	5	3	5	6	7	6	4	5
28	6	3	5	3	5	3	5	0	0	3	3	0	0	5	0
29	4	3	4	3	3	0	0	0	0	4	4	0	0	5	0
30	4	6	5	6	9	4	10	3	1	3	3	2	2	7	3
31	5	5	4	7	8	4	10	3	2	5	5	3	4	8	3
32	3	3	5	7	7	9	10	3	2	5	3	7	5	5	2
33	2	3	5	7	7	9	10	3	2	2	3	6	4	5	2
34	3	4	6	4	3	3	8	1	1	3	3	3	2	5	2
35	6	7	4	3	3	0	9	0	1	0	2	3	1	5	3
36	9	8	5	5	6	6	8	2	2	2	4	5	6	6	3
37	4	9	6	4	10	8	8	9	1	3	9	7	5	3	2
38	4	9	6	6	9	9	7	9	1	2	10	8	5	5	2
39	10	6	9	10	9	10	10	10	10	10	8	10	10	10	10
40	10	6	9	10	9	10	10	10	10	10	10	10	10	10	10
41	10	7	8	0	2	1	2	0	10	2	0	3	0	0	10
42	10	3	8	0	1	1	0	0	10	0	0	0	0	0	10
43	3	4	9	8	2	4	5	3	6	2	1	3	3	3	8
44	7	7	7	6	9	8	8	6	8	8	10	8	8	6	5
45	9	6	10	9	7	7	10	2	1	5	5	7	8	4	5
46	9	8	10	10	7	9	10	3	1	5	7	9	9	4	4
47	0	7	10	3	5	0	10	0	0	2	2	0	0	0	0
48	0	6	10	1	5	0	10	0	0	2	2	0	0	0	0

Example 3.1

Table 3.1 shows the scores on 15 variables of 48 applicants for a certain post. The 15 variables were as follows:

(1) Form of letter of application   (9) Experience
(2) Appearance   (10) Drive
(3) Academic ability   (11) Ambition
(4) Likeability   (12) Grasp
(5) Self-confidence   (13) Potential
(6) Lucidity   (14) Keenness to join
(7) Honesty   (15) Suitability
(8) Salesmanship

One of the questions of interest here is how the variables cluster, in the sense that some of the qualities may be correlated or confused in the judge's mind. (There was no purpose in clustering the candidates – only one was to be chosen.) Table 3.2 gives the correlation matrix of the data.

Table 3.2   *Correlation matrix of the data of Table 3.1*

	1	2	3	4	5	6	7	8	9	10	11	12	13	14	15
1	1·00	·24	·04	·31	·09	·23	−·11	·27	·55	·35	·28	·34	·37	·47	·59
2		1·00	·12	·38	·43	·37	·35	·48	·14	·34	·55	·51	·51	·28	·38
3			1·00	·00	·00	·08	−·03	·05	·27	·09	·04	·20	·29	−·32	·14
4				1·00	·30	·48	·65	·35	·14	·39	·35	·50	·61	·69	·33
5					1·00	·81	·41	·82	·02	·70	·84	·72	·67	·48	·25
6						1·00	·36	·83	·15	·70	·76	·88	·78	·53	·42
7							1·00	·23	−·16	·28	·21	·39	·42	·45	·00
8								1·00	·23	·81	·86	·77	·73	·55	·55
9									1·00	·34	·20	·30	·35	·21	·69
10										1·00	·78	·71	·79	·61	·62
11											1·00	·78	·77	·55	·43
12												1·00	·88	·55	·53
13													1·00	·54	·57
14														1·00	·40
15															1·00

The clustering can be carried out by inspection on a matrix of this size. We will, as a first step, decide that two vectors are "close" if their correlation is ⩾0·7 in absolute value, and that a set of vectors form a cluster if all their correlations exceed that amount. Looking through the table, we find that there are two cases (6, 12) and (12, 13) which have the greatest correlation, 0·88. Let us take the pair (12, 13) as the nucleus of a possible cluster. We then observe that variables 10 and 11 are also highly correlated with both and with each other, so we can add them to the cluster, obtaining the set (10, 11, 12, 13). Further inspection shows that 6 and 8 can also be added, and 5 can be adjoined if we condone the correlation between 5 and 13,

namely 0·67, as a slight shortfall on our critical value. Thus we cluster the
seven variables

$$(5, 6, 8, 10, 11, 12, 13) \tag{3.1}$$

No other variables can be added. If we delete from the matrix the seven
rows and columns corresponding to the variables of (3.1) we get for the remaining eight Table 3.3. No correlation now attains 0·7 and we should, on
this criterion, say that the other variables do not cluster, or that they provide 8 clusters of each. There is scope for personal choice in the order
in which variables are put together in this way, but the final result should
be the same.

Table 3.3   *Correlation matrix after clustering of variables*

	1	2	3	4	7	9	14	15
1	1·00	·24	·04	·31	−·11	·55	·47	·59
2		1·00	·12	·38	·35	·14	·28	·38
3			1·00	·00	−·03	·27	−·32	·14
4				1·00	·65	·14	·69	·33
7					1·00	−·16	·45	·00
9						1·00	·21	·69
14							1·00	·40
15								1·00

The dividing line of 0·7 is somewhat arbitrary. If we wish to relax the
requirement for closeness we might group together (1, 9, 15), with an average correlation of 0·61, and (4, 7, 14) with an average of 0·60. Variable 3
(Academic Ability) stands alone, one observes with mixed feelings. Variable
2 (Appearance) is fairly highly correlated with 11, 12, 13 and might perhaps
be included in the main group.

Provisionally, then, we may consider the hypothesis that the variables are
by no means independent, and that judgement is being exercised in four
dimensions: the first, a blend of Appearance, Self-confidence, Lucidity, Drive,
Ambition, Grasp, and Potential (perhaps facets of the candidate's ability to
project an outlooking personality); the second, Letter of Application, Experience, and Suitability (perhaps an experience component); the third, Likeability, Honesty, and Keenness to Join (frankness or likeability); the fourth,
Academic Ability. At this point we can say nothing about the relative importance of these components, if they exist, except that Suitability is most
highly correlated with the Experience group. We shall consider the data again
in a later chapter from the point of view of Factor Analysis.

3.7   If the relations among individuals could be handled in the same way
as those among variables there would be little more to say, at least about
continuous variation. The method of selecting a nucleus and building up

around it has the great advantage that we can see what we are doing and even get some intuitive feeling for the "shape" of the cluster. The size of n, however, makes the approach very time-consuming. Among 1000 members there are nearly half a million distances between pairs, and although a search routine could doubtless be set up on an electronic machine, such procedures are well known to require more time than can usually be afforded.

3.8   There is, however, a prior difficulty to be overcome. In general we cannot set up a scale-free measure of the distance between two members. (Attempts based on the so-called "Mahalanobis distance", which we shall discuss in Chapter 10, fail, for reasons there indicated.) It will be clear on a little reflection that if we calculate the correlation between the variate-values borne by a pair of objects, it is not invariant under a change in the scale of one of them. The point will be clear from the consideration of a cluster in two dimensions on variables, say x_1 and x_2. If we begin with a simple cluster – looking, for example, like the spots of the number 5 on the face of a die – it is plain that if we hold the scale of x_1 constant but stretch the scale of x_2 we can separate the points as much as we like. It seems essential, then, that in clustering individuals we must impose some sort of scale on prior grounds. I shall assume that all the variables have been scaled to unit variance. This constraint can be relaxed if we have reason to suppose that some variables are more important than others, provided that we can quantify their relative importance. For instance, if x_1 is thought to be twice as important as x_2, we can assign it a variance one-quarter of that of x_2.

3.9   Whatever method of clustering is employed, we require some measure of "distance" between two individuals. This need not necessarily be an ordinary Euclidean distance but it should obey certain criteria. In particular it is desirable, as a general rule, that the distance from A to B is the same as that from B to A. This is not quite such a trivial requirement as it sounds; some sociological groupings are based on affinity or liking between persons, and it is not necessary that B likes A if A likes B. A further desirability of the distance metric, though again not a necessary one, is that the triangular inequality be obeyed, namely that the distance from A to B is not greater than the sum of the distances of A to C and C to B.

3.10   The distance metric which I prefer is the simple Euclidean distance. If the variables are x_1 to x_p, we define the square of the distance from object A to object B (which I shall call the "deviance") as

$$d^2(A,B) = \sum_{i=1}^{p} (x_{iA} - x_{iB})^2. \tag{3.2}$$

Geometrically the deviance is the square on the hypotenuse in p dimensions and corresponds to what one ordinarily thinks of as distance in the metrical sense. This, however, is not the main reason for using it.

(1) In the first place, d^2 is invariant under a rotation of the axes and in particular is the same if we use principal component scores instead of the original variables. If then the complex is such as to collapse into a lower space, or to have a number of eigenvalues which are very small, we can approximate to the deviance by the use of fewer dimensions. This is a valuable feature where a large number of intercorrelated variables are concerned and can greatly reduce the arithmetic. For example, in a study of 157 British towns, Moser and Scott (1961) measured 57 variables. It turned out, however, that 95 percent of the variation was accounted for by five principal components. If one is content to discard the extra 5 percent as unimportant for the classificatory process, the exploration can be reduced from 57 dimensions to five.

(2) In delimiting clusters we are usually interested in the average deviance among a subset of m points, not the individual $\frac{1}{2}m(m-1)$ deviances, just as, in Example 3.1, we were concerned with the average correlation within a group. Now it so happens that the mean deviance of the set can be determined without the calculation of all individual deviances. If the ith variable on the jth member is x_{ij}, we have for the mean deviance of a set of m

$$\frac{1}{m(m-1)} \sum_{i=1}^{p} \sum_{j=1}^{m} \sum_{k=1}^{m} (x_{ij} - x_{ik})^2$$

$$= \frac{1}{m(m-1)} \sum_{i=1}^{p} \sum_{j=1}^{m} \sum_{k=1}^{m} \{(x_{ij} - x_{i.}) - (x_{ik} - x_{i.})\}^2,$$

where $x_{i.}$ is the mean of x_i over the m members,

$$= \frac{1}{m(m-1)} \sum_{i=1}^{p} \left\{ \sum_{j} \sum_{k} (x_{ij} - x_{i.})^2 + \sum_{j} \sum_{k} (x_{ik} - x_{i.})^2 \right.$$
$$\left. - 2 \sum_{j} \sum_{k} (x_{ij} - x_{i.})(x_{ik} - x_{i.}) \right\}.$$

The cross-product term vanishes and the other two are equal. Thus

$$\text{Average deviance} = \frac{2}{m-1} \sum_{j=1}^{m} \sum_{i=1}^{p} (x_{ij} - x_{i.})^2. \tag{3.3}$$

Instead, then, of having to calculate the $\frac{1}{2}m(m-1)$ deviances we have only to calculate the m deviances from the centre of gravity. The order of calculation is reduced from $\frac{1}{2}m^2$ to $p + m$ (p for the centre of gravity and m for the deviances from it). Again the saving in arithmetic is large.

Example 3.2

Although we can sidestep the calculation of individual deviances, it is worth considering whether any insight is to be gained into the nature of the clustering by forming a frequency distribution of the deviances. One might suppose that separation into clusters would throw up some kind of multi-modality. Unfortunately this does not appear to be so.

Consider, for example, two clusters of n_1 and n_2 members which are well separated, so that the deviances within clusters are small and those between clusters are large. We shall then have

$$\tfrac{1}{2}n_1(n_1 - 1) + \tfrac{1}{2}n_2(n_2 - 1)$$

small deviances and $n_1 n_2$ large ones. A frequency distribution might be bimodal, but the relative frequencies are very dependent on the proportion of n_1 to n_2. The difference, in fact, is

$$\tfrac{1}{2}n_1(n_1 - 1) + \tfrac{1}{2}n_2(n_2 - 1) - n_1 n_2 = \tfrac{1}{2}(n_1 - n_2)^2 - \tfrac{1}{2}(n_1 + n_2), \quad (3.4)$$

which will be positive if n_1 and n_2 are fairly different, but negative if they are equal. If now we suppose that the distribution of deviances in the two clusters has a substantial dispersion, as compared to the deviances between them, almost any shape of frequency distribution might result; and *a fortiori* for more than two clusters.

3.11   When the distance metric has been decided upon we can begin the search for clusters in several ways. One is the build-up around a nucleus which, as already mentioned, may be prohibitively time-consuming. Another is to split the group into two, and then each group into further groups and so on. The third, which is the one we shall adopt, is to start by dividing the group into subsets on an arbitrary basis and then to improve the grouping iteratively by a computer program. The procedure is as follows.

(1)   We decide on some number q, say 20, as an upper limit to the number of clusters in which we are interested.

(2)   We determine in the p-dimensional space q cluster centres in some arbitrary way, say by spacing them at intervals. These initially chosen points are merely for starting off on an iterative process.

(3)   The observations are considered one at a time and allocated to the nearest centre. We thus define 20 "clusters".

(4)   The centre of gravity of each cluster is computed.

(5)   The observations are considered one at a time. An observation is moved to another cluster (or rather, the boundaries of the cluster are moved to include that point) if such a move reduces the sum of the deviances from the observations to their cluster centres *when the centres of the two clusters are simultaneously moved to their new centres of gravity.*

If an observation is moved in this way, the new c.g.'s are computed and replace the former ones.

(6)   The set of observations is re-scanned until no observation is moved to a new cluster. At this point we pause, having arrived at q clusters in such a way that the sum of deviances from the centres of gravity cannot be lowered by moving an observation from one cluster to another.

3.12   This procedure is reasonable, but it may not arrive at a set of clusters such that the overall sum of deviances from the c.g.'s is an absolute minimum. It is conceivable that the set of clusters so obtained may depend on the starting points and the order in which the observations are considered one by one. However, if there is any serious doubt, the process can be repeated with a different set of starting points. In practice we are usually working towards a smaller set of clusters than 20, and when amalgamation takes place, as described below, it usually happens that even when different initial sets of 20 are chosen, they converge to agreement by the time the process has merged clusters to give, say, 10.

3.13   At the next stage we look for a set of $q - 1$ clusters. To provide a suitable starting point, a pair of clusters is chosen for merging. The basis for selection is that their combination should minimize the increment in the deviances from their cluster means. When we merge two clusters some increment is inevitable. We consider all pairs of clusters and choose the one which has the smallest. This, however, does not involve permanent merging of the two; it is only an expedient for starting with $q - 1$ clusters in an efficient way. The programme of computation then returns to the beginning and reallocates the observations to $q - 1$ clusters in the manner described in paragraph **3.11**. The calculation then proceeds to $q - 2$, $q - 3$, etc., clusters in turn. At each stage we compute the cluster centres, the distances between cluster centres, the sums of deviances of observations from their respective cluster centres, and finally print out the individuals in each cluster.

3.14   We thus arrive at a series of answers to the problem of defining clusters, with $q, q - 1, q - 2, \ldots$, clusters. The question now is, which do we choose? A criterion suggested by E.M.L. Beale (but there may be others) is based on an analogy with the analysis of variance, though we must be careful to remember that it is an analogy only, not a formal equivalence. For c clusters we add together the deviances from the respective cluster means. This may be regarded as a residual sum of squares, and we denote it by R_c. If it were zero, all the clusters would be condensed at their centres of gravity. If it is large, the points are widely dispersed and the clustering, if any, is very loose.

Think of the observations in the First Space as contained in a space of

Table 3.4 *Multiple measurements on three varieties of Iris*

Iris setosa				Iris versicolor				Iris virginica			
Sepal length	Sepal width	Petal length	Petal width	Sepal length	Sepal width	Petal length	Petal width	Sepal length	Sepal width	Petal length	Petal width
5·1	3·5	1·4	0·2	7·0	3·2	4·7	1·4	6·3	3·3	6·0	2·5
4·9	3·0	1·4	0·2	6·4	3·2	4·5	1·5	5·8	2·7	5·1	1·9
4·7	3·2	1·3	0·2	6·9	3·1	4·9	1·5	7·1	3·0	5·9	2·1
4·6	3·1	1·5	0·2	5·5	2·3	4·0	1·3	6·3	2·9	5·6	1·8
5·0	3·6	1·4	0·2	6·5	2·8	4·6	1·5	6·5	3·0	5·8	2·2
5·4	3·9	1·7	0·4	5·7	2·8	4·5	1·3	7·6	3·0	6·6	2·1
4·6	3·4	1·4	0·3	6·3	3·3	4·7	1·6	4·9	2·5	4·5	1·7
5·0	3·4	1·5	0·2	4·9	2·4	3·3	1·0	7·3	2·9	6·3	1·8
4·4	2·9	1·4	0·2	6·6	2·9	4·6	1·3	6·7	2·5	5·8	1·8
4·9	3·1	1·5	0·1	5·2	2·7	3·9	1·4	7·2	3·6	6·1	2·5
5·4	3·7	1·5	0·2	5·0	2·0	3·5	1·0	6·5	3·2	5·1	2·0
4·8	3·4	1·6	0·2	5·9	3·0	4·2	1·5	6·4	2·7	5·3	1·9
4·8	3·0	1·4	0·1	6·0	2·2	4·0	1·0	6·8	3·0	5·5	2·1
4·3	3·0	1·1	0·1	6·1	2·9	4·7	1·4	5·7	2·5	5·0	2·0
5·8	4·0	1·2	0·2	5·6	2·9	3·6	1·3	5·8	2·8	5·1	2·4
5·7	4·4	1·5	0·4	6·7	3·1	4·4	1·4	6·4	3·2	5·3	2·3
5·4	3·9	1·3	0·4	5·6	3·0	4·5	1·5	6·5	3·0	5·5	1·8
5·1	3·5	1·4	0·3	5·8	2·7	4·1	1·0	7·7	3·8	6·7	2·2
5·7	3·8	1·7	0·3	6·2	2·2	4·5	1·5	7·7	2·6	6·9	2·3
5·1	3·8	1·5	0·3	5·6	2·5	3·9	1·1	6·0	2·2	5·0	1·5
5·4	3·4	1·7	0·2	5·9	3·2	4·8	1·8	6·9	3·2	5·7	2·3
5·1	3·7	1·5	0·4	6·1	2·8	4·0	1·3	5·6	2·8	4·9	2·0
4·6	3·6	1·0	0·2	6·3	2·5	4·9	1·5	7·7	2·8	6·7	2·0
5·1	3·3	1·7	0·5	6·1	2·8	4·7	1·2	6·3	2·7	4·9	1·8
4·8	3·4	1·9	0·2	6·4	2·9	4·3	1·3	6·7	3·3	5·7	2·1
5·0	3·0	1·6	0·2	6·6	3·0	4·4	1·4	7·2	3·2	6·0	1·8
5·0	3·4	1·6	0·4	6·8	2·8	4·8	1·4	6·2	2·8	4·8	1·8
5·2	3·5	1·5	0·2	6·7	3·0	5·0	1·7	6·1	3·0	4·9	1·8
5·2	3·4	1·4	0·2	6·0	2·9	4·5	1·5	6·4	2·8	5·6	2·1
4·7	3·2	1·6	0·2	5·7	2·6	3·5	1·0	7·2	3·0	5·8	1·6
4·8	3·1	1·6	0·2	5·5	2·4	3·8	1·1	7·4	2·8	6·1	1·9
5·4	3·4	1·5	0·4	5·5	2·4	3·7	1·0	7·9	3·8	6·4	2·0
5·2	4·1	1·5	0·1	5·8	2·7	3·9	1·2	6·4	2·8	5·6	2·2
5·5	4·2	1·4	0·2	6·0	2·7	5·1	1·6	6·3	2·8	5·1	1·5
4·9	3·1	1·5	0·2	5·4	3·0	4·5	1·5	6·1	2·6	5·6	1·4
5·0	3·2	1·2	0·2	6·0	3·4	4·5	1·6	7·7	3·0	6·1	2·3
5·5	3·5	1·3	0·2	6·7	3·1	4·7	1·5	6·3	3·4	5·6	2·4
4·9	3·6	1·4	0·1	6·3	2·3	4·4	1·3	6·4	3·1	5·5	1·8
4·4	3·0	1·3	0·2	5·6	3·0	4·1	1·3	6·0	3·0	4·8	1·8
5·1	3·4	1·5	0·2	5·5	2·5	4·0	1·3	6·9	3·1	5·4	2·1
5·0	3·5	1·3	0·3	5·5	2·6	4·4	1·2	6·7	3·1	5·6	2·4
4·5	2·3	1·3	0·3	6·1	3·0	4·6	1·4	6·9	3·1	5·1	2·3
4·4	3·2	1·3	0·2	5·8	2·6	4·0	1·2	5·8	2·7	5·1	1·9
5·0	3·5	1·6	0·6	5·0	2·3	3·3	1·0	6·8	3·2	5·9	2·3
5·1	3·8	1·9	0·4	5·6	2·7	4·2	1·3	6·7	3·3	5·7	2·5
4·8	3·0	1·4	0·3	5·7	3·0	4·2	1·2	6·7	3·0	5·2	2·3
5·1	3·8	1·6	0·2	5·7	2·9	4·2	1·3	6·3	2·5	5·0	1·9
4·6	3·2	1·4	0·2	6·2	2·9	4·3	1·3	6·5	3·0	5·2	2·0
5·3	3·7	1·5	0·2	5·1	2·5	3·0	1·1	6·2	3·4	5·4	2·3
5·0	3·3	1·4	0·2	5·7	2·8	4·1	1·3	5·9	3·0	5·1	1·8

volume V. If the clusters divide this space into regions of about the same content and σ_c^2 denotes the mean deviance of points from their centre of gravity, the value of σ_c^2 will decrease as c increases according to the formula

$$c\sigma_c^p = k_0, \tag{3.5}$$

where k_0 does not depend on c. This implies that

$$\sigma_c^2 = kc^{-2/p}, \tag{3.6}$$

where k is some constant independent of c. Then, for approximately spherical clusters in which the variables are independent,

$$E(R_c) = k(n-c)c^{-2/p}, \tag{3.7}$$

n being the sample number.

Consider now two clusterings of c_1 and c_2, $c_2 > c_1$. We must then have $R_{c_1} > R_{c_2}$. Again approximately, the expected increase in R as we condense from c_2 to c_1 can be represented as

$$E\left(\frac{R_{c_1} - R_{c_2}}{R_{c_2}}\right) = \frac{n-c_1}{n-c_2}\left(\frac{c_2}{c_1}\right)^{2/p} - 1. \tag{3.8}$$

We regard this as a variance-increase and compute the F-ratio

$$F(c_1, c_2) = \frac{R_{c_1} - R_{c_2}}{R_{c_2}} \bigg/ \left\{\frac{n-c_1}{n-c_2}\left(\frac{c_2}{c_1}\right)^{2/p} - 1\right\}, \tag{3.9}$$

with $p(c_2 - c_1)$ and $p(n - c_2)$ degrees of freedom. If for any given c_1 it is significant for any c_2, we say that the representation in terms of c_1 is not entirely adequate.

This is a highly heuristic criterion and must not be applied automatically, especially if the clusters are not globular. It does, however, for the want of anything better, give us some guide as to the appropriate point at which to stop clustering. We must then inspect the clusters to see if the results are acceptable on general grounds.

Example 3.3

In Table 3.4 we reproduce some well-known data cited by Fisher (1936) in his famous paper on discrimination. We happen to know here that there are three different groups, and if they were distinguishable on the four variables (petal length, petal width, sepal length, and sepal width) we should expect a cluster analysis to separate them. An analysis was therefore conducted on all 150 members ($n = 150$, $p = 4$) considered as a single group. The process was started with six clusters and run down to one cluster, namely the whole set. The following F-ratios were obtained, the figures in brackets being the significance levels:

	1	2	3	4	5
2	8·19 (0·0%)				
3	10·11 (0·0%)	4·00 (0·3%)			
4	10·48 (0·0%)	3·83 (0·0%)	2·32 (5·5%)		
5	9·77 (0·0%)	3·35 (0·0%)	1·89 (5·9%)	1·18 (31·5%)	
6	8·67 (0·0%)	2·82 (0·0%)	1·48 (12·7%)	0·83 (58·1%)	0·95 (77·6%)

$$(3.10)$$

The table is to be read from south-east to north-west. Starting with six clusters we see that in comparison with five, at 77·6 percent significance, nothing is lost by going to five clusters. The bottom row indicates that we can go to three clusters, but not to two, before the clusters are adequately separated. This is confirmed by the second row from the bottom. We can go from 5 to 4 (31·5 percent significance) and to 3 (5·9 percent) but not to 2. The third row from the bottom and the other two rows confirm the conclusion that three clusters are required.

We then examine these clusters. The following are the distances between them:

Cluster number	Cluster size	Distance from grand mean	Sepal length	Sepal width	Petal length	Petal width
			Cluster centred at			
1	62	0·747	5·902	2·748	4·394	1·434
2	50	2·649	5·006	3·428	1·462	0·246
3	38	2·390	6·850	3·074	5·742	2·071
Grand mean			5·843	3·057	3·758	1·199
Standard deviation			0·828	0·436	1·765	0·762

Distances between cluster centres

From cluster number	To cluster number		
	1	2	3
1	0·000		
2	3·357	0·000	
3	1·797	5·018	0·000

Cluster 2 is well separated from the others. The average distance of this group of 50 from its centre is 0·482. A print-out of the distances of individuals (which we omit to save space) shows that no member is more than 1·248 from the centre, and all are more than 3 from the next nearest cluster. It so happens, in fact, that this group of 50 consists entirely of *setosa*.

Clusters 1 and 3 are closer together. The average distance of the members of cluster 1 from the centre is 0·738 and the maximum 1·661. For cluster 3 the corresponding figures are 0·720 and 1·530. In fact, the 62 members of cluster 1 comprise 48 *versicolor* and 14 *virginica*; cluster 3 comprises 2 *versicolor* and 36 *virginica*. When we consider the data from the point of view of discrimination (Chapter 10) we shall find, in fact, that there is some overlap between the varieties. Our clustering process, of course, does not allow of overlapping clusters, and there has been some misclassification, which is an essential feature of the data.

3.15 When the variables are not continuous some difficulties arise in defining what we mean by a cluster. Consider an extreme case in which there are three variables, each of them a simple dichotomy. We can represent them as (0, 1) variables on the corners of a cube in a space of type formally equivalent to the First Space. There are, of course, clusters at each corner of the cube, but such clustering is rather trivial; it consists of placing together those individuals which are identical in every respect; or, to put it another way, the "clusters" are simply the members in the cells of the 2 × 2 × 2 table which defines the data.

Example 3.4

Suppose that we had a set of 45 persons each of whom was charged before a court with burglary. Attributes which we might observe on them are A, whether married (a = not married); B whether previously convicted (b = not previously convicted); C whether armed (c = not armed). Let the frequencies be

$$(A\ B\ C) = 16, \quad (A\ B\ c) = 4, \quad (a\ B\ C) = 8, \quad (A\ b\ C) = 8,$$
$$(A\ b\ c) = 2, \quad (a\ B\ c) = 2, \quad (a\ b\ C) = 4, \quad (a\ b\ c) = 1$$

These frequencies are chosen so that the attributes are completely independent; for example,

$$\frac{16}{45} = \frac{(A\ B\ C)}{45} = \frac{(A)}{45}\frac{(B)}{45}\frac{(C)}{45} = \frac{30}{45}\frac{30}{45}\frac{36}{45}.$$

Thus, each variable contributes independently to the clustering, if any. Suppose now that we decided to cluster those members which have two attributes in common. There are several ways in which we could group together the members with two common letters (either capital or small). In the cube which represents the situation spatially, any two vertices which are joined by an edge to another vertex have just two common letters. But we have no clue as to whether to join, say $(A\ B\ C)$ and $(A\ B\ c)$ or $(A\ B\ C)$ and $(A\ b\ C)$; and, if we adjoin all three, the groups $(A\ B\ c)$ and $(A\ b\ C)$ have only one letter in common. The problem is indeterminate.

3.16 Where grouping or clustering are concerned, the simple dichotomy into presence or absence of a variable is not a symmetrical one, in the sense that presence of the attributes may be more significant of likeness than its absence. Given any two individuals, we can always find more respects in which they differ than in which they are the same, so that any measure of similarity can be attenuated practically to zero by including enough attributes. Nevertheless, methods of clustering have been developed based on attributes. If two individuals have m attributes in common out of p, a so-called "similarity" index can be defined as m/p. The complementary quantity $1 - m/p$ can be regarded as a distance function. It is zero if the two individuals are identical in every observed respect, unity if they are different in every observed respect. Clustering can proceed accordingly by putting together those individuals with small "distances" as so defined. It will, however, be evident from Example 3.4 that this is a rather hazardous procedure and may give misleading results.

3.17 It is not easy to recommend what procedure should be followed with dichotomous data regardless of the circumstances of the case, and indeed it may well be that "clustering" is not legitimate unless there is some well-defined purpose which dictates its nature. The first thing to decide is whether it is the presence or absence of the attribute which is of importance. For example, with the 45 burglars of Example 3.4, if we are concerned with the severity of the offence, the presence of previous convictions and the carrying of arms are the important features — as it were the positive ends of the polarity. Marital status may have little to do with the point at issue, and it would be better to keep the married and unmarried groups together.

3.18 Intermediate between continuous variation and dichotomy there lies the case when the variables are ordered into classes. For example, we may have individuals recorded according to eight income groups, five social classes, three classes of academic degree, and so on. Geometrically the individuals, instead of being on the corners of a hypercube, lie on the nodes of a lattice. In such a case similar considerations arise to those in the extreme case of dichotomy, but in a sense the difficulties are not so severe. Very often we can quantify the intervals between the classes and represent, say, a five-way polytomy by attaching the variable values $-2, -1, 0, 1, 2$ on the assumption that the "distances" between classes are equal. The continuous variable can, of course, be regarded as an extreme case when the class intervals diminish to zero. Again some care is necessary because a "cluster" would be an aggregate of neighbouring lattice points, not a central condensation of the continuous kind.

3.19 In some instances the score-values to be attached to a polytomy can be assigned from considerations of ranking. If there is reason to believe that the observed classification is based on a more-or-less continuous variable which, for some reason or other, is not itself measured, we may regard the polytomy as a heavily tied ranking.

Example 3.5

Suppose we have 100 individuals classified by social grade as follows:

A	B	C	D	E
5	15	30	40	10

If this were a ranking of 100 people, we might regard the A group as ranked 1 to 5 and allot them the mean rank $\frac{1}{5}(1 + 2 + 3 + 4 + 5) = 3$. The next group would be a tied ranking of 15, from 6 to 20, with a mean rank of 13. The third, fourth and fifth groups have mean ranks 35·5, 60·5, 95·5.

It must be remembered that if we are to standardize to unit variance, these scores also need reduction. It may be shown (Kendall, 1971) that the variance of a set of n tied into groups of length $t_1, t_2, \ldots, t_k$ is

$$\frac{1}{12n} \left\{ n^3 - n - \sum_{i=1}^{k} (t_i^3 - t_i) \right\}, \qquad (3.11)$$

so that separate calculation of the variance is not necessary.

NOTES

(1)   The statistical theory (if such it can be called) of cluster analysis is relatively new and may develop quite substantial new features over the coming years. In particular, the use of visual display units attached to computers and the possibility of using light-pens to modify clusters on the spot may have a great impact on practical work once the requisite equipment becomes inexpensive.

(2)   The problem of defining the "shape" of a cluster in p dimensions is not an easy one. The distribution of distances of the points from their centre of gravity will give some idea whether there is any central clustering (as against a toroid, where there would be no small distances). The determination of the convex hull of a set requires linear programming, and the maximum-to-minimum distance between apexes of the hull will give some idea of the elongation of the cluster.

(3)   When clusters are well defined, it is of some interest to show which variables are contributing most to the separation. (In a study of TV broadcasts we were surprised to find cricket and religion together in a cluster, the explanation being that they were both considered as very restful to watch.) Sometimes an inspection of the means of variables is suggestive; for instance, in Example 3.3 petal length and petal width are obviously more effective than the sepal measurements. However, it may be that some variables are the operating separators for Clusters A and B, whereas others are so for B and C and C and A. To examine all the possibilities for more than two or three variables is usually too costly in machine time. Another mode of approach, once the clusters

are determined, is to set up discriminant functions (Chapter 10) and to consider the coefficients in the discriminator.

(4)   If the variables are highly correlated it is probably best to work in terms of principal component scores (which should *not* be reduced to unit variance). If, for example, x_1 were almost a linear function of x_2 and x_3, its inclusion in the distance function might be equivalent to counting x_2 and x_3 twice.

(5)   The heuristic way in which we approach clustering implies that, even if the members are a random sample, it is difficult to express the results probabilistically in terms of standard errors or confidence regions. A sample of moderate size from, say, a spherically symmetric distribution such as the joint distribution of a set of independent normal standardized variables, will quite possibly produce clusters by sheer chance. One useful technique, if the number n is large enough, is to split the set at random into two or more subsets and to see how far the clusters in the subsets are similar.

(6)   A case which has been relatively little studied, but is of some importance in ecology and medicine, is that wherein there is a background density of population from which the observed points emanate. For example, the cases of incidence of some disease will cluster in more densely populated areas, simply because there are more individuals at risk in them. To remove this effect and consider the clustering which may be due to other causes we have to weight the distance metric.

Suppose the whole domain consists of sub-areas, the mean density of the whole domain is n, and that of the jth sub-area is n_j. A suitable distance function might then be of the form

$$\frac{1}{n^2} \sum_{i=1}^{p} (x_{iA} - x_{iB})^2 n_j n_k,$$

where A occurs in sub-area j and B in sub-area k. If all the n_j are equal, this is the ordinary deviance. Within a district ($j = k$) it weights the distance according to the relative density in that district. The "distance" between a point in district j and one in district k may be regarded as the geometric mean of what it would have been if both points were in j or in k. Other weightings are possible, but the above preserves to some extent the relation between average deviances of pairs of points and their average deviance from the centre of gravity.

(7)   Many other methods of clustering have been proposed. Some proceed by continual partitioning, some insist on a hierarchical classification, some look for clumping or multimodality in the variables. A good test of such procedures is whether they consume an inordinate amount of computer time. Reference may be made to the monograph by Brian Everett (1974), *Cluster Analysis*, published for the Social Science Research Council by Heinemann.

4

Factor analysis

4.1 In 1904 Charles Spearman published a famous paper which may be regarded as the starting point of factor analysis. Spearman was working with scores obtained in examinations, and noticed certain systematic effects in the matrix of correlations between scores. One of his examples, obtained from 33 "high-class preparatory school" children, was as follows:

	1	2	3	4	5	6
1 Classics	—	·83	·78	·70	·66	·63
2 French	·83	—	·67	·67	·65	·57
3 English	·78	·67	—	·64	·54	·51
4 Mathematics	·70	·67	·64	—	·45	·51
5 Discrimination	·66	·65	·54	·45	—	·40
6 Music	·63	·58	·51	·51	·40	—

(4.1)

The subjects are deliberately arranged so that the correlation attenuates from left to right. It is then noteworthy that the values decrease in each row, more or less to the same extent. Spearman pointed out that this effect could be accounted for if the score on the ith variable were composed of two parts:

$$x_i = a_i f + \epsilon_i, \qquad (4.2)$$

where f is a random variable common to all the x's and ϵ_i is specific to x_i. The concept was that performance on any test was a sum of a general factor (identifiable with general intelligence) and an ability specific to that particular test. If f and the ϵ_i are all independent,

$$\mathrm{cov}\,(x_i, x_j) = \mathrm{E}\,(a_i f + \epsilon_i)(a_j f + \epsilon_j)$$
$$= a_i a_j \,\mathrm{var}\, f,$$

so that

$$\frac{\mathrm{cov}\,(x_i, x_j)}{\mathrm{cov}\,(x_i, x_k)} = \frac{a_j}{a_k}, \text{ independent of } i,$$

which would account for the proportionality observed.

4.2   It was found that a good many batteries of tests gave a more complicated structure than could be accounted for by the simple model (4.2), and a natural generalization was to suppose that performance was the sum of a number of different factors. Specifically we suppose that there are $m < p$ factors $\zeta_1, \zeta_2, \ldots, \zeta_m$ and that

$$x_i = \sum_{k=1}^{m} l_{ik}\zeta_k + \epsilon_i. \tag{4.3}$$

The l's are not now (as in component analysis) the coefficients in an orthogonal rotation. They measure, regardless of the individual member, the extent to which a variable x is compounded of the underlying factors ζ. The score of any particular individual j is regarded as the selection of a value of each ζ_k peculiar to him and a value of the ϵ also specific to him:

$$x_{ij} = \sum_{k=1}^{m} l_{ik}\zeta_{kj} + \epsilon_{ij}. \tag{4.4}$$

The model structure assumes further

(1)   that the ζ's are independent normal variables with zero mean and unit variance;
(2)   that each ϵ is independent of all other ϵ's and of the ζ's,
(3)   and consequently that the covariance matrix of the ζ's is the identity matrix $\mathbf{I}$,
(4)   and that the covariance matrix of the ϵ's is a diagonal matrix Σ with elements which we write as σ_i^2.

4.3   This is quite a different model from the one which we have considered in principal component analysis. The latter, in fact, can hardly be described as a model — it is merely a convenient variate-transformation. If m were equal to p, of course, the ϵ's would be unnecessary and we should revert to the case of a transformation to a set of independent variables. But the point of factor analysis is that m is less than p, and is to be chosen as small as possible so as to present as economical a model as possible.

4.4   Furthermore, since the ζ's are assumed to be independent normal variables with unit variance, they are not uniquely determinable, and in consequence nor are the l's. In virtue of our assumptions, we have

$$\text{cov}\,(x_i, x_j) = \mathrm{E}\left\{\left(\sum_{k=1}^{m} l_{ik}\zeta_k + \epsilon_i\right)\left(\sum_{t=1}^{m} l_{jt}\zeta_t + \epsilon_j\right)\right\}$$

$$= \sum_{k=1}^{m} l_{ik}l_{jk}, \quad i \neq j \tag{4.5}$$

$$\text{var}\, x_i = \sum_{k=1}^{m} l_{ik}^2 + \sigma_i^2. \tag{4.6}$$

The relations may be summarized in matrix form. If l is a $p \times m$ matrix the model is

$$x = l\zeta + \epsilon, \tag{4.7}$$

and equations (4.5) and (4.6) are equivalent to

$$c = ll' + \Sigma, \tag{4.8}$$

where c is the covariance matrix of the x's.

If we consider a non-singular orthogonal transformation of the ζ's to new variables η, say by

$$\eta = M\zeta, \tag{4.9}$$

then the η's will also be standardized normal independent variables. In place of (4.8) we should find that

$$\begin{aligned} c &= E\,(lM\zeta + \epsilon)(lM\zeta + \epsilon)' \\ &= E\,(lM\zeta\zeta'M'l') + \Sigma \\ &= ll' + \Sigma. \end{aligned} \tag{4.10}$$

Thus, as we remarked, the model is indeterminate within a rotation of the factors. We may regard it as determining the *factor space*, within which we can determine any orthogonal set of axes we like to correspond to the factors.

4.5   Here also we stand in contrast to the case of principal components. In the latter we find preferred transformations giving stationary values to the resulting ξ's. Here all the ζ's have the same unit variance. On the other hand, the l's in principal component analysis are a unitary orthogonal set. In factor analysis they are neither unitary nor necessarily orthogonal. Furthermore, we note that whereas we can transform from x's to ξ's and vice versa, we cannot express the ζ's in terms of x's. The matrix l, being $p \times m$, is not invertible. The best we can hope to do is to make the problem determinate by fixing the axes in the factor space, and then estimate the l's and Σ, so expressing the structure of the x's in terms of the ζ's.

4.6   The more usual statistical word to denote the coefficients l in component analysis would be "weight". For historical reasons, mostly in psychological work, the coefficients in (4.3) are called "loadings", l_{ik} being the loading of the ith variable on the kth factor or the loading of the kth factor in the ith variable. The variables ϵ are referred to either as "specific factors" or as "residuals". From (4.6) we see that the variance of x_1 is the sum of a component in the l's and the specific σ_i^2. The first part is called the "communality" and the second the "residual variance".

4.7   In equation (4.8) there are pm constants l and p constants in Σ,

$p(m + 1)$ in all. There are $\frac{1}{2}p(p + 1)$ variances and covariances. Thus the parameters are not determinate if

$$\frac{1}{2}p(p + 1) < p(m + 1) \tag{4.11}$$

or $m > \frac{1}{2}(p - 1)$, and even if this inequality were reversed, would only be determinate within a rotation in the m-space. If we impose conditions on the l's to fix one of the possible sets of axes in the m-space we impose $\frac{1}{2}m(m - 1)$ further conditions. Then the situation is indeterminate if

$$\frac{1}{2}p(p + 1) + \frac{1}{2}m(m - 1) \leqslant p(m + 1) \tag{4.12}$$

or equivalently, if

$$(p - m)^2 \leqslant p + m, \tag{4.13}$$

which may also be expressed as

$$(p + \tfrac{1}{2} - m)^2 \leqslant \tfrac{1}{4}(8p + 1). \tag{4.14}$$

Example 4.1

Let us consider a few simple cases, first of all $p = 2$, $m = 1$. The model is then

$$x_1 = l_{11}\zeta + \epsilon_1$$
$$x_2 = l_{21}\zeta + \epsilon_2.$$

We have the three equations

$$\operatorname{var} x_1 = l_{11}^2 + \sigma_1^2$$
$$\operatorname{var} x_2 = l_{21}^2 + \sigma_2^2$$
$$\operatorname{cov}(x_1, x_2) = l_{11}l_{21}.$$

Here are three equations in four unknowns. Some extra condition is required to make the problem determinate, e.g. some assumption about the ratio of σ_1^2 to σ_2^2.

With $p = 3$, $m = 2$ we should have, as the reader may easily verify, six equations in nine unknowns. With $p = 3$, $m = 1$ we have six equations in six unknowns. With $p = 5$, $m = 1$ we have 15 equations in 10 unknowns and hence one factor is not, in general, enough; but with $m = 2$ there are 15 unknowns and the problem is resolvable so far as concerns (4.11). If we impose $\frac{1}{2}m(m - 1)$ further conditions, (4.13) is obeyed for $p = 5$, $m = 2$, and there are more than enough equations so that some reconciliation, e.g. by least squares, is required.

We have, however, to be careful in arguments relating number of variables to number of equations. Consider the case of five variables and two factors:

$$x_1 = l_1\zeta_1 + m_1\zeta_2 + \epsilon_1$$
$$\cdot \qquad \cdot \qquad \cdot \qquad \cdot$$
$$x_5 = l_5\zeta_1 + m_5\zeta_2 + \epsilon_5.$$

There are five equations giving the variances of the ϵ's in terms of the other parameters. For the l's we have 10 equations typified by

$$l_1 l_2 + m_1 m_2 = \text{cov}(x_1, x_2).$$

But these are not enough to determine the l's and m's, within a rotation. For example, if

$$l_1 = \lambda_1 \cos \theta + \mu_1 \sin \theta$$

$$m_1 = \lambda_1 \sin \theta - \mu_1 \cos \theta, \quad \text{etc.},$$

$$l_1 l_2 + m_1 m_2 = (\lambda_1 \cos \theta + \mu_1 \sin \theta)(\lambda_2 \cos \theta + \mu_2 \sin \theta) +$$

$$+ (\lambda_1 \sin \theta - \mu_1 \cos \theta)(\lambda_2 \sin \theta - \mu_2 \cos \theta)$$

$$= \lambda_1 \lambda_2 + \mu_1 \mu_2.$$

Lawley and Maxwell (1971) point out that in the special case where there is an equality in (4.13) there may or may not be an acceptable solution. For example, with $p = 3$ and $m = 1$ and the matrix

$$
\begin{array}{ccc}
1 & 0\cdot84 & 0\cdot60 \\
 & 1 & 0\cdot35 \\
 & & 1
\end{array}
$$

the equations $l_1 l_2 = 0\cdot84$, $l_2 l_3 = 0\cdot35$, $l_3 l_1 = 0\cdot60$ yield $l_1^2 = 1\cdot44$, which would give $\sigma_1^2 = \text{var } \epsilon_1 = 1 - 1\cdot44$, a negative value.

4.8 In the past, psychologists used various devices to simplify the determination of the factor loadings. Some of these devices involved guessing the residual variances or the communalities; others involved plausible but not necessarily optimal procedures. The advent of the computer has rendered much of this work obsolete. In this chapter we shall notice two methods of procedure, although the theory of the second, based on maximum likelihood methods of estimation, will have to await attention until Chapter 6.

Let us write

$$c_a = c - \Sigma, \tag{4.15}$$

so that the dispersion matrix is "adjusted" by subtracting σ_j^2 from the variance of x_j. Then

$$c_a = ll'. \tag{4.16}$$

To make the situation determinate we now impose the condition that

$$l'l = \Lambda_m, \quad \text{a diagonal matrix.} \tag{4.17}$$

Then

$$c_a l = ll'l = l\Lambda_m. \tag{4.18}$$

This bears a strong resemblance to the fundamental equation of Chapter 2 giving principal components; but the diagonals of the covariance matrix are now the communalities, which we write as h_i^2, given by

$$h_i^2 = \text{var } x_i - \sigma_i^2, \tag{4.19}$$

and Λ is only an $m \times m$ diagonal matrix. As before, we find for the first column of products in (4.18)

$$(h_1^2 - \lambda_1)l_{11} + \qquad c_{12}l_{21} + \ldots + c_{1p}l_{p1} = 0$$

$$c_{21}l_{11} + (h_2^2 - \lambda_1)l_{21} + \ldots + c_{2p}l_{p1} = 0$$

$$\cdot \quad \cdot \quad \cdot \quad \cdot \quad \cdots \quad \cdot \quad \cdot$$

$$c_{m1}l_{11} + \qquad c_{m2}l_{21} + \ldots + (h_m^2 - \lambda_1)l_{m1} + c_{mp}l_{p1} = 0$$

$$\cdot \quad \cdot \quad \cdot \quad \cdot \quad \cdots \quad \cdot \quad \cdot$$

$$c_{p1}l_{11} + \qquad c_{p2}l_{21} + \ldots + (h_p^2 - \lambda_1)l_{p1} = 0,$$

leading to

$$|\mathbf{c}_a - \lambda \mathbf{I}| = 0. \tag{4.20}$$

In contradistinction to the principal components case, we require only m of the eigenvalues and the corresponding l's. We choose the m largest eigenvalues. It follows from (4.17) that these values have the largest sums $\sum\limits_{i=1}^{m} l_{ij}^2$ and hence the greatest factor loadings (in this average sum of squares sense), and hence they correspond to the most "important" factors. If we have chosen the communalities correctly there will be m non-negative roots in λ. If we have chosen them badly the matrix may cease to be positive definite and we must reduce m or re-estimate the communalities.

4.9   The unfortunate feature of this approach is that we do not know the communalities. Two practical approaches seem to work reasonably well:

(1)   We may do a component analysis (i.e. with known communalities) and come to the conclusion that only m eigenvalues are important, the others being discardable. If we standardize the retained m ξ-variables and express the x's in terms of these as ζ variables, the other $p - m$ variables which are neglected explicitly can be regarded as providing the error terms.

(2)   We may, in the manner of Lawley and Maxwell (1971) treat the problem as one of maximizing likelihood in a multivariate complex, and estimate the l's and σ^2's simultaneously. This requires a rather different normalizing procedure from that of (4.17). The resulting equations are somewhat complicated and iterative methods of solution were not always successful. Jöreskog (1971) has had greater success by approaching the maximum directly, and a machine program is now available. Compare **6.22**.

Example 4.2

We revert to the data of Table 3.1 (scores of 48 applicants on 15 variables). The correlation matrix was given in Table 3.2, and since all the variables were scored on the same scale we may work with correlations instead of covariances. The following are the eigenvalues of the correlation matrix:

	Eigenvalue	Cumulative proportion		Eigenvalue	Cumulative proportion
1	7·50	0·500	9	0·26	0·958
2	2·06	0·637	10	0·20	0·972
3	1·46	0·735	11	0·15	0·982
4	1·21	0·815	12	0·09	0·988
5	0·74	0·864	13	0·08	0·993
6	0·49	0·897	14	0·06	0·998
7	0·35	0·921	15	0·03	1·000
8	0·31	0·941			

The first four components account for 81·5 percent of the variation; these are the components with eigenvalues greater than unity. More components can be taken if desired — the choice is somewhat arbitrary and in practice several choices may have to be compared — but four will illustrate the method of approach.

The loadings on the factors were as in Table 4.1. It must be remembered that these are not the direction cosines of the corresponding eigenvectors but those quantities multiplied by the square root of the corresponding eigenvalues, because we are working with factors which have unit variance. It should also be noted that if we read this table across, so that, e.g.,

$$x_1 = 0.445\zeta_1 + 0.618\zeta_2 + 0.372\zeta_3 - 0.119\zeta_4 + \epsilon_1,$$

we are, in effect, defining the residual ϵ_1 and hence the communality. It is more customary to read the table downwards and examine which variables are most heavily weighted in the factors. We must, again, remember that we have only estimated the factor space of four dimensions, and that these particular factors are arbitrary within a rotation.

Taking the factors as they stand, we find a somewhat confusing picture. Factor 1 seems loaded in all the variables except, perhaps, x_3. Factor 2 is loaded in x_1, x_7, x_9 and x_{15}. Factor 3 is loaded most heavily in x_3, x_4 and x_{14}. Factor 4 is loaded on x_3 and perhaps x_7.

4.10 We are still free to rotate the factors if we wish, that is to say, to find orthogonal linear transformations to four new factors. It is natural to

Table 4.1　　*Loadings of the first four eigenvectors of Table 3.2*

Variable	Factors				Name of variable
	1	2	3	4	
1	0·445	0·618	0·372	−0·119	Letter
2	0·583	−0·048	−0·017	0·289	Appearance
3	0·109	0·340	−0·500	0·710	Academic
4	0·616	−0·180	0·575	0·361	Likeability
5	0·799	−0·358	−0·295	−0·178	Self-confidence
6	0·865	−0·188	−0·182	−0·070	Lucidity
7	0·433	−0·576	0·361	0·448	Honesty
8	0·881	−0·056	−0·245	−0·230	Salesmanship
9	0·365	0·795	0·099	0·070	Experience
10	0·864	0·066	−0·100	−0·165	Drive
11	0·873	−0·098	−0·256	−0·206	Ambition
12	0·908	−0·031	−0·134	0·092	Grasp
13	0·912	0·035	−0·078	0·213	Potential
14	0·710	−0·114	0·560	−0·234	Keenness
15	0·646	0·605	0·103	−0·028	Suitability

do so by bringing the new factors into some sort of accordance with the variables, for example, by reducing as many loadings as possible to zero, and to maximize as many of the others as possible. Furthermore, if the factors are independent, it is also desirable for a particular variable not to be heavily loaded on more than one factor. Various methods have been proposed for the purpose. A convenient one is the Varimax method (Kaiser, 1958). The factors are rotated in such a way as to maximize the sum of the variances of the squared loadings within each column of the rotated loading matrix. If h_i^2 is the communality, equal to $\sum_{k=1}^{m} l_{ik}^2$, we maximize, for new l's, the expression

$$\sum_{j=1}^{m}\left[p \sum_{i=1}^{p} \left\{\frac{l_{ij}^2}{h_i^2}\right\}^4 - \left\{\sum_{i=1}^{p} \left(\frac{l_{ij}^2}{h_i^2}\right)\right\}^2 \right].\qquad(4.21)$$

The general effect is to produce large or small loadings and to avoid intermediate sizes. There is no reason why the loadings should be less than unity — they are not direction cosines as in the case of principal components.

Other methods (quartimax, quartimin) have been proposed. They are all directed more or less to the same end.

Example 4.3

Table 4.2 gives a varimax rotation on the data of Table 4.1. So far as concerns interpretation the position is now somewhat clearer. The fourth

Table 4.2   *Rotated factors from Table 4.1*

Variable	Rotated factors 1	2	3	4	Name of variable
1	0·116	0·830	0·109	−0·136	Letter
2	0·436	0·152	0·401	0·228	Appearance
3	0·062	0·128	0·007	0·928	Academic
4	0·216	0·245	0·872	−0·081	Likeability
5	0·918	−0·104	0·166	−0·063	Self-confidence
6	0·863	0·099	0·259	0·004	Lucidity
7	0·216	−0·242	0·863	0·001	Honesty
8	0·917	0·206	0·087	−0·051	Salesmanship
9	0·083	0·852	−0·052	0·212	Experience
10	0·798	0·352	0·161	−0·049	Drive
11	0·917	0·162	0·106	−0·038	Ambition
12	0·806	0·257	0·338	0·146	Grasp
13	0·741	0·329	0·420	0·227	Potential
14	0·437	0·364·	0·538	−0·522	Keenness
15	0·381	0·797	0·078	0·084	Suitability

factor seems to be entirely one of academic ability, which is not heavily loaded in any of the other variables except negatively on keenness. Factor 3 is loaded most heavily in likeability and honesty − it could be interpreted as a general likeability of person, apart from particular qualifications. Factor 2, if we must christen it, seems to be one of experience. Factor 1 is the usual blend of qualities such as self-confidence, salesmanship, ambition, and lucidity. No variable is heavily loaded in more than one factor. Two variables, appearance and keenness, are not very heavily loaded in any factor.

It is interesting to compare this analysis with the one which we obtained in Example 3.1. The conclusions are much the same. The basis of approach, however, is different. In this example we have sought four *independent* factors and, by rotation, have provisionally identified them with groups of variables. In Example 3.1 we grouped the variables, but had we then identified those groups with factors, we should have derived *dependent* factors.

4.11   The factor analyst sometimes admits non-orthogonal factors and hence modifies the original hypothesis of equation (4.3) in an important respect. The mathematics then become even more complicated. It appears to me that if dependent factors are to be admitted, one should proceed directly to a cluster analysis of the variables and bypass the procedure requiring factor estimation and rotation. Sometimes, as in Example 3.1, the grouping is enough. If we wish to go further and identify loadings on the oblique (dependent, non-orthogonal factors), we can proceed as follows.

A cluster analysis is performed on the variables and forced to produce as many clusters as the m factors required. Within each cluster an average vector is determined. In the Second Space this is equivalent to putting a vector through the centre of gravity of the ends of the unit vectors defining the cluster. The m principal component axes are then transformed so that the new axes go through these centroids. The coefficients of this transformation, combined with those of the transformation to principal components, give the loadings on the oblique factors.

Table 4.3   *Oblique factor loadings of data in Table 3.1*

Variable	Oblique factors				Name of variable
	1	2	3	4	
1	−0·120	0·888	0·108	−0·161	Letter
2	0·332	−0·007	0·343	0·295	Appearance
3	−0·000	−0·000	−0·000	0·939	Academic
4	−0·183	0·167	1·021	0·167	Likeability
5	1·123	−0·810	−0·150	−0·147	Self-confidence
6	0·954	−0·099	−0·006	−0·048	Lucidity
7	−0·058	−0·426	1·002	0·272	Honesty
8	1·074	0·039	−0·242	−0·177	Salesmanship
9	−0·181	0·893	−0·076	0·146	Experience
10	0·850	0·211	−0·100	−0·138	Drive
11	1·077	−0·014	−0·219	−0·154	Ambition
12	0·796	0·055	0·122	0·126	Grasp
13	0·654	0·124	0·252	0·244	Potential
14	0·236	0·309	0·511	−0·439	Keenness
15	0·231	0·774	−0·032	0·015	Suitability

Example 4.4

For comparison with previous examples, Table 4.3 gives the loadings on four oblique factors determined in the foregoing manner. Again the same general conclusions emerge, but in somewhat stronger form. The sizes of the major loadings in the four factors are roughly comparable, which would be interpreted as implying that if the variable values are really being allocated, more or less subconsciously, on the factors of a four-dimensional complex, those factors are about equally important.

4.12   In **4.9** (2) we referred to the machine program developed by Jöreskog for estimating the coefficients and testing for the number of factors. Later, in **6.22**, we examine the method by which, on the assumption of normality in the variables, we arrive at expressions of the likelihood in terms of the

Table 4.4 *Maximum likelihood estimates of the factor loadings for the data of Table 4.1, four factors fitted, unrotated matrix*

Variable	Factor loadings				Name of variable
	1	2	3	4	
1	0·460	−0·136	0·528	−0·226	Letter
2	0·450	0·295	0·062	0·133	Appearance
3	−0·109	0·521	0·386	0·257	Academic
4	0·695	−0·204	0·136	0·514	Likeability
5	0·716	0·464	−0·396	−0·072	Self-confidence
6	0·773	0·440	−0·123	0·049	Lucidity
7	0·480	−0·078	−0·253	0·615	Honesty
8	0·785	0·439	−0·099	−0·227	Salesmanship
9	0·280	0·097	0·697	−0·234	Experience
10	0·800	0·314	0·085	−0·161	Drive
11	0·783	0·441	−0·135	−0·179	Ambition
12	0·797	0·439	0·083	0·115	Grasp
13	0·797	0·440	0·210	0·221	Potential
14	0·908	−0·391	−0·022	−0·035	Keenness
15	0·560	0·155	0·596	−0·242	Suitability

Table 4.5 *Maximum likelihood estimates of the factor loadings for the data of Table 4.1, four factors fitted, varimax rotation*

Variable	Factor loadings				Name of variable
	1	2	3	4	
1	0·126	−0·102	0·722	0·113	Letter
2	0·453	0·166	0·141	0·242	Appearance
3	0·072	0·692	0·121	−0·009	Academic
4	0·226	−0·051	0·243	0·834	Likeability
5	0·922	−0·089	−0·096	0·147	Self-confidence
6	0·842	0·054	0·119	0·287	Lucidity
7	0·249	−0·024	−0·223	0·753	Honesty
8	0·896	−0·067	0·238	0·074	Salesmanship
9	0·092	0·188	0·763	−0·056	Experience
10	0·762	−0·048	0·394	0·180	Drive
11	0·898	−0·059	0·190	0·110	Ambition
12	0·783	0·168	0·281	0·358	Grasp
13	0·725	0·274	0·351	0·445	Potential
14	0·419	−0·569	0·406	0·561	Keenness
15	0·357	0·098	0·781	0·059	Suitability

parameters of interest. A solution of the equations derived by differentiating this likelihood in order to maximize it leads to practical difficulties of a somewhat obscure nature. Jöreskog's program seeks iteratively for the actual maximum more directly by a hill-climbing technique. It first fits one factor and rotates to a varimax solution; it then tests on a χ^2 whether one factor is adequate. If not, it proceeds to two factors, and so on.

Example 4.5

The program was run on the data of our previous examples. For comparison, Tables 4.4 and 4.5 give the results as far as four factors.

The program, however, announces that four factors are not an adequate fit, giving a value of 103·9 for χ^2 with 51 degrees of freedom. We proceed to 7 factors before a satisfactory fit is reached. The loadings on a 7-factor analysis with varimax rotation are as in Table 4.6.

Table 4.6   *Maximum likelihood estimation of the factor loadings for the data of Table 4.1, seven factors fitted, varimax rotation*

Variable	Factor loadings						
	1	2	3	4	5	6	7
1	0·129	0·074	0·665	−0·096	0·017	−0·042	0·267
2	0·329	0·242	0·182	0·095	0·611	−0·013	−0·006
3	0·048	−0·017	0·097	0·688	0·043	0·007	0·008
4	0·249	0·759	0·252	−0·058	0·090	−0·096	0·204
5	0·882	0·184	−0·082	−0·074	0·190	0·059	−0·045
6	0·907	0·266	0·136	0·046	−0·042	−0·290	−0·016
7	0·199	0·911	−0·224	−0·013	0·174	−0·094	−0·204
8	0·875	0·082	0·264	−0·076	0·140	0·043	−0·058
9	0·073	−0·027	0·718	0·158	0·069	0·036	0·009
10	0·780	0·197	0·386	0·026	−0·051	0·398	−0·023
11	0·874	0·036	0·157	−0·052	0·382	0·142	0·205
12	0·775	0·346	0·286	0·172	0·143	−0·159	0·111
13	0·703	0·409	0·354	0·329	0·140	0·070	0·193
14	0·432	0·540	0·381	−0·540	−0·013	0·099	0·275
15	0·313	0·079	0·909	0·049	0·142	0·027	−0·214

A comparison with the results which we obtained earlier has to be made with some care because the order of the factors by importance is not necessarily the same for the two methods. However, the results of the 7-factor analysis seem to confirm those we found by cluster analysis in Example 3.1. The first factor is heavily loaded on variables 5, 6, 8, 10, 11, 12 and 13. Variables 2 (factor 5) and 3 (factor 4) appear to be isolated. Of the remainder we might group together 1, 9, 15 (factor 3) and 4, 7, 14 (factor 2). The last two factors contribute very little.

Further experience is necessary before one can make definite recommendations as to preference, but in the present state of knowledge it seems to me that the method of clustering is adequate for most purposes.

NOTES

(1) In the estimation of l and Σ by the method of 4.8 it might be thought that some iterative procedure was possible. For example, having made a provisional estimate of Σ from the residuals of m principal components, one might go back to (4.21), derive a second estimate of the l's, then a second estimate of Σ, and so on. Unfortunately a routine sometimes fails to converge or, if it does, requires hundreds of iterations. The reasons are obscure, but there is enough experience to render the use of iterative routines of this type rather hazardous.

(2) In this chapter the question as to "how many factors" has been answered on a somewhat intuitive basis. A more refined criterion will be discussed in Chapter 6.

(3) In principal components we can invert the relation between x and ξ to give the ξ's as linear functions of the x's and hence derive the values of ξ for any one individual — the component scores. In factor analysis such a thing is impossible. We cannot determine factor scores, the $p \times m$ matrix of l's being non-invertible. In my opinion we have to regard factor scores as *theoretically* unmeasurable. It seems to me preferable not to attempt the impossible. However, two methods of "estimating" have been proposed, namely Thomson's (1939) and Bartlett's (1948). Thomson assumed that the factors may be regressed on the p variables so that the factor scores f are given by

$$\mathbf{f} = l'c^{-1}\mathbf{x},$$

which is tantamount to assuming that $\mathbf{fx}'$, the covariance matrix of factor scores and variables, is estimated by the loading matrix l'.

Bartlett *defines* the factor scores as having a minimal sum of squares of standardized residuals, i.e. he minimizes

$$\sum_{i=1}^{p} (x_i - \sum_j l_{ij} f_j)^2 / \text{var } x_i.$$

5

Canonical correlations

5.1   Principal component analysis and factor analysis are concerned with the interdependence of a group of variables. We now extend such work to the interdependence between two groups of variables. Suppose we have one set of variables $x_1, \ldots, x_p$ and another set $y_1, \ldots, y_q$, where $p \leqslant q$. (Since the relationships are symmetrical we lose no generality by this assumption.) In line with the general policy of transforming the variables so as to simplify the structure, we now seek for new variables $\xi_1, \ldots, \xi_p, \eta_1, \ldots, \eta_q$, such that the correlations between members of the ξ group and members of the η group either vanish or have stationary values. We shall in fact show that there exists a transformation such that

(1)   All the ξ's and all the η's have zero mean and unit variance.

(2)   Any ξ in the p-group is uncorrelated with any other ξ in the p-group; and any η in the q-group is uncorrelated with any other η in the q-group.

(3)   The correlations between the ξ's and η's all vanish except for p correlations $\rho_1, \rho_2, \ldots, \rho_p$, which may be taken to be the correlations between ξ_1 and η_1, ξ_2 and $\eta_2, \ldots, \xi_p$ and η_p.

(4) ρ_1 is the greatest correlation that can be found between a linear function of the x's and a linear function of the y's; ρ_2 is the greatest correlation between linear functions of x and y which are uncorrelated with those contributing to ρ_1; and so on.

5.2   In our approach to principal components we began by looking for uncorrelated variables and then proved that the ones we found had stationary properties. In the following we shall go the other way round, looking for stationary correlations and then proving that the resulting variables are uncorrelated. The resulting variables are said to be in *canonical form* and the correlations are called *canonical correlations*. In contrast to principal components, the canonical correlation components do not have stationary variances represented by the eigenvalues. The optimization is concerned with the reduction of correlation to a minimal set.

5.3   We will suppose our x's and y's to be standardized to zero mean and unit variance, and their covariances typified by c_{jk}. It will be convenient to distinguish the dispersions in the p group by Greek suffixes and those in the q-group by Roman suffixes, so that a covariance between a member of the p-group and a member of the q-group might be $c_{\alpha j}$ with one Greek and one Roman suffix. We then seek transformations

$$\xi_i = \sum_{\alpha=1}^{p} l_{i\alpha} x_\alpha, \quad i = 1, 2, \ldots, p; \tag{5.1}$$

$$\eta_j = \sum_{a=1}^{q} m_{ja} y_a, \quad j = 1, 2, \ldots, q. \tag{5.2}$$

If their variances are unity, we have

$$\sum_{\alpha, \beta=1}^{p} l_{i\alpha} l_{i\beta} c_{\alpha\beta} = 1, \quad i = 1, 2, \ldots, p; \tag{5.3}$$

$$\sum_{a, b=1}^{q} m_{ja} m_{jb} c_{ab} = 1, \quad j = 1, 2, \ldots, q. \tag{5.4}$$

Consider one pair, ξ_i, η_i. We require their correlation R_i to be stationary for variations in l and m, i.e. we require

$$R = \sum_{\alpha=1}^{p} \sum_{a=1}^{q} l_{i\alpha} m_{ia} c_{\alpha a} \tag{5.5}$$

to be stationary, subject to (5.3) and to (5.4) with j equal to i. Take two undetermined multipliers $\frac{1}{2}\lambda$ and $\frac{1}{2}\mu$. Omitting the suffix i for convenience, we then require an unconditioned minimum of

$$\sum_{\alpha, a} l_\alpha m_a c_{\alpha a} - \frac{1}{2}\lambda \left(\sum_{\alpha, \beta} l_\alpha l_\beta c_{\alpha\beta}^{-1} \right) - \frac{1}{2}\mu \left(\sum_{a, b} m_a m_b c_{ab} - 1 \right) \tag{5.6}$$

On differentiation with respect to m and also with respect to l this leads to

$$\sum_\alpha l_\alpha c_{\alpha a} - \mu \sum_b m_b c_{ab} = 0. \tag{5.7}$$

$$\sum_a m_a c_{\alpha a} - \lambda \sum_\beta l_\beta c_{\alpha\beta} = 0. \tag{5.8}$$

Multiply the first equation by m_a and sum. In virtue of (5.4) and (5.5) we find

$$R = \mu.$$

Similarly, from the second equation,

$$R = \lambda,$$

so that

$$\lambda = \mu = R. \tag{5.9}$$

Equations (5.7) and (5.8) are then solvable for l and m if their determinant vanishes, that is to say if the $(p + q) \times (p + q)$ determinant

$$\begin{vmatrix} -\lambda c_{\alpha\beta} & c_{\alpha b} \\ c_{a\beta} & -\lambda c_{ab} \end{vmatrix} = 0. \qquad (5.10)$$

In full this would appear as

$$\begin{vmatrix} -\lambda c_{11} & -\lambda c_{12} & \ldots -\lambda c_{1p} & c_{1,p+1} & \cdots & c_{1,p+q} \\ -\lambda c_{21} & -\lambda c_{22} & \ldots -\lambda c_{2p} & c_{2,p+1} & \cdots & c_{2,p+q} \\ \cdot & \cdot & \cdot & \cdot & \cdot & \cdot \\ -\lambda c_{p1} & -\lambda c_{p2} & \ldots -\lambda c_{p,p} & c_{p,p+1} & \cdots & c_{p,p+q} \\ c_{p+1,1} & c_{p+1,2} \cdots & c_{p+1,p} & -\lambda c_{p+1,p+1} & \cdots -\lambda c_{p+1,p+q} \\ c_{p+2,1} & c_{p+2,2} \cdots & c_{p+2,p} & -\lambda c_{p+2,p+1} & \cdots -\lambda c_{p+2,p+q} \\ \cdot & \cdot & \cdot & \cdot & \cdots & \cdot \\ c_{p+q,1} & c_{p+q,2} \cdots & c_{p+q,p} & -\lambda c_{p+q,p+1} & \cdots -\lambda c_{p+q,p+q} \end{vmatrix} = 0.$$

$$(5.11)$$

This repulsive expression can be substantially reduced. Multiplying the first p rows by $-\lambda$ and dividing the last q columns by $-\lambda$, we reduce (5.10) to

$$(-\lambda)^{q-p} \begin{vmatrix} \lambda^2 c_{\alpha\beta} & c_{\alpha b} \\ c_{a\beta} & c_{ab} \end{vmatrix} = 0. \qquad (5.12)$$

Now premultiply by the $(p+q) \times (p+q)$ determinant

$$\begin{vmatrix} I_p & -c_{\alpha b} c_{ab}^{-1} \\ 0 & c_{ab}^{-1} \end{vmatrix}. \qquad (5.13)$$

We then find

$$(-\lambda)^{q-p} \begin{vmatrix} \lambda^2 c_{\alpha\beta} - c_{\alpha b} c_{ab}^{-1} c_{a\beta} & c_{\alpha b} - c_{\alpha b} c_{ab}^{-1} c_{ab} \\ c_{ab}^{-1} c_{a\beta} & I_q \end{vmatrix} = 0,$$

or, since the North-East corner is zero,

$$(-\lambda)^{q-p} |\lambda^2 c_{\alpha\beta} - c_{\alpha b} c_{ab}^{-1} c_{a\beta}| = 0. \qquad (5.14)$$

This is now only a $p \times p$ determinant. Apart from the zero values of λ, it will have p roots in λ^2. From (5.9) we see that these are the squares of the stationary values of the correlations, which are thus determined from (5.14) except for sign.

5.4   Given the λ's, the corresponding l's and m's can be determined. It remains to show that the ξ's and η's are uncorrelated among and between themselves, except for the correlations ρ_i which can be taken to be the correlations between ξ_i and η_i.

For any i we have, from (5.7) and (5.8),

$$\sum_\alpha l_{i\alpha} c_{\alpha a} = \rho_i \sum_b m_{ib} c_{ab} \qquad (5.15)$$

$$\sum_a m_{ia} c_{\alpha a} = \rho_i \sum_\beta l_{i\beta} c_{\alpha\beta}. \qquad (5.16)$$

There will be a similar pair for some other suffix j. Between the four variables ξ_i, ξ_j, η_i, η_j there are six correlations. It will be enough to prove that four vanish, namely

$$E(\xi_i \xi_j) = \sum_{\alpha,\beta} l_{i\alpha} l_{j\beta} c_{\alpha\beta}, \qquad E(\xi_j \eta_i) = \sum_{\alpha,b} l_{j\alpha} m_{ib} c_{\alpha b}. \qquad (5.17)$$

$$E(\xi_i \eta_j) = \sum_{\alpha,b} l_{i\alpha} m_{jb} c_{\alpha b}, \qquad E(\eta_i \eta_j) = \sum_{a,b} m_{ia} m_{jb} c_{ab}, \qquad (5.18)$$

Multiply (5.15) by m_{ja} and sum. In virtue of (5.17) and (5.18) we have

$$E(\xi_i \eta_j) = \rho_i E(\eta_i \eta_j). \qquad (5.19)$$

Likewise from (5.16) multiplied by $l_{j\alpha}$ and summed we find

$$E(\xi_j \eta_i) = \rho_i E(\xi_i \xi_j). \qquad (5.20)$$

Interchanging i and j, we find from (5.19) and (5.20)

$$\rho_i E(\eta_i \eta_j) = \rho_j E(\xi_i \xi_j), \qquad (5.21)$$

and interchanging i, j in this,

$$\rho_j E(\eta_i \eta_j) = \rho_i E(\xi_i \xi_j). \qquad (5.22)$$

It follows from (5.21) and (5.22) that unless $\rho_i^2 = \rho_j^2$,

$$E(\eta_i \eta_j) = E(\xi_i \xi_j) = 0, \qquad (5.23)$$

which was to be proved. In a similar way it may be shown that the other covariances vanish.

There is a degenerate case if two canonical roots are equal. We may then choose our l's and m's to obey certain additional conditions, and in particular ensure that

$$E(\xi_i \xi_j) + E(\eta_i \eta_j) = 0. \qquad (5.24)$$

It will follow from (5.22) that each expectation vanishes unless $\rho_i = \rho_j = 0$. Even in this case (5.19) and (5.20) show that the two expectations vanish.

5.5 The main result is then established. We can transform the x's and the y's to a situation in which the whole correlation structure is summarized into p canonical correlations. We also know that no linear function of x can be more highly correlated with a linear function of y than ρ_1. The main problem lies in the interpretation of the results. It is, in one sense, a more intensified version of the problem of interpreting principal components or factors. We note with regret that one resource open to us in that case, the rotation of factors or transformation to dependent components, is not available here because any further modification of the canonical variables would disturb the canonical correlations and reintroduce further correlations into the system.

5.6   In (5.14) the determinant embodies terms which all depend on λ^2. By subtracting suitable multiples of rows and columns this can, however, be reduced to one in which λ^2 occurs only down the diagonals, and the solution then becomes very similar to that required for the determination of eigenvalues. The general remarks made in Chapter 2 about degenerate solutions apply here – the occurrence of c_{ab}^{-1} in (5.14) warns us to be careful of collinearities in the y's. We could, in fact, take equation (5.14) one step further to obtain

$$| \lambda^2 I_p - c_{\alpha\beta}^{-1} c_{\alpha b} c_{ab}^{-1} c_{a\beta} | = 0, \qquad (5.25)$$

which not only exhibits the eigenvalue nature of the solution but also underlines the dangers arising from degeneracy in $c_{\alpha\beta}$ or c_{ab}.

Example 5.1 (Hotelling, 1936)

140 seventh-grade schoolchildren received four tests on (a) reading speed, (b) reading power, (c) arithmetic speed, and (d) arithmetic power. The correlations in performance were as follows (figures to four places for greater accuracy):

$$\begin{bmatrix} 1\cdot0 & 0\cdot6328 & 0\cdot2412 & 0\cdot0586 \\ & 1\cdot0 & -0\cdot0553 & 0\cdot0655 \\ & & 1\cdot0 & 0\cdot4248 \\ & & & 1\cdot0 \end{bmatrix} \qquad (5.26)$$

The point of interest here is not so much the correlations among tests as whether reading ability (as measured on two tests) is correlated with arithmetic ability (similarly measured).

The determinant in one of the forms (5.14) or (5.25) is then easily solvable to yield

$$\lambda = 0\cdot3945 \quad \text{or} \quad 0\cdot0689. \qquad (5.27)$$

The lower root seems of no importance. The larger indicates some relationship, not a very strong one. We can, if necessary, find the values of l's and m's. For instance, with the larger root, from (5.7) and (5.8),

$$\left. \begin{array}{r} l_1 + 0\cdot6328\,l_2 - 0\cdot6114\,m_1 - 0\cdot1485\,m_2 = 0 \\ 0\cdot6328\,l_1 + l_2 + 0\cdot1402\,m_1 - 0\cdot1660\,m_2 = 0 \\ -0\cdot6114\,l_1 + 0\cdot1402\,l_2 + m_1 + 0\cdot4248\,m_2 = 0 \\ -0\cdot1485\,l_1 - 0\cdot1660\,l_2 + 0\cdot4248\,m_1 + m_2 = 0. \end{array} \right\} \qquad (5.28)$$

One of these equations is linearly dependent on the other three, and we can solve only for the ratios of l's and m's. We find

$$l_1 : l_2 : m_1 : m_2 = -2{\cdot}7772 : 2{\cdot}2655 : -2{\cdot}4404 : 1.$$

Thus the transformed variables are

$$k_1\xi_1 = -2{\cdot}7772x_1 + 2{\cdot}2655x_2 \qquad (5.29)$$
$$k_2\eta_1 = -2{\cdot}4404x_3 + x_4, \qquad (5.30)$$

where k_1 and k_2 can be adjusted, if required, so that ξ_1 and η_1 have unit variance.

One would not draw very firm conclusions from data such as these, but it is worth remarking that although x_1 and x_2 are quite highly correlated in (5.26), they occur with opposite signs in (5.29); and so for x_3 and x_4 in (5.30). One might, perhaps, regard this as suggestive of the fact that ρ_1 compares the speed–power difference in reading with that in arithmetic, rather than indicating a blend of the two skills.

Example 5.2 (F.V. Waugh, 1942)

Waugh took, for each of the years 1921 to 1940 inclusive, the prices of beef steers and hogs and the *per capita* consumption of beef and pork (excluding lard) for the USA. The prices were "deflated" by dividing by an index of *per capita* income, that is to say they purport to measure the changes in price relative to a stable value of money, and are given as dollars per 100 lb at Chicago. The consumption is given in pounds per annum.

We thus have, for 20 years ($n = 20$), a multivariate situation with $p = 2$, $q = 2$. We require to discuss the question how far meat consumption and meat prices are related, "meat" for this purpose including beef and pork but not mutton or chicken or minor sources of meat.

This, in one way, is a simplified form of the problem of canonical correlation analysis because effectively we need only one linear combination of the price — and consumption — variables and the greatest canonical correlation. The others are of minor interest. The correlation matrix was

	X_1	X_2	X_3	X_4
X_1 (steer prices)	1·0	0·181 26	−0·563 96	−0·498 98
X_2 (hog prices)		1·0	0·354 94	−0·756 71
X_3 (beef consumption)			1·0	−0·102 93
X_4 (pork consumption)				1·0

Let us note that these correlations make economic sense. The correlations between steer prices and beef consumption and between hog prices and pork consumption are negative; a rise in price means a fall in consumption. But the correlation between hog price and beef consumption is positive; when pork goes up in price there is a switch to beef, the consumption of which also goes up. (But the correlation between steer prices and pork consumption is negative, so that substitutional effects are not entirely straightforward.)

The classical way of discussing the question would probably have been to form an index-number (a weighted average) of prices and another (also a weighted average) of consumption and to investigate the relation between the two. The weights used in constructing these indices would have to be selected on prior grounds of a somewhat arbitrary kind: and the resulting correlation would, of course, depend on them. If we adopt the standpoint of canonical analysis, the weights are determined for us by the condition that the correlation is a maximum. But we must always remember that there is nothing in the economics of the situation to compel the supposition that correlations are maximized.

The greatest canonical correlation in this present example was $-0{\cdot}846\,66$ (greatest, that is, in absolute value). We choose the negative sign in the light of the economics of the situation — an increase in price should lead to a decrease in consumption. The new variables, in terms of the standardized X's, were

$$\xi_1 = \text{constant } (52{\cdot}62X_1 + 47{\cdot}38X_2)$$
$$\eta_1 = \text{constant } (25{\cdot}38X_3 + 74{\cdot}62X_4), \tag{5.31}$$

where we have chosen the weights so as to add up to 100.

On looking at these values we see that the signs at least are acceptable. ξ_1 is an average of prices with nearly equal weights, a reasonable average for prices which both relate to the same quantity of meat; and the weights in η_1 are also both positive. Whether they are "reasonable" in the sense of providing a good index-number of the consumption of meat is another question, and one which has no answer unless we can say for what purposes the index is intended.

Example 5.3 (F.V. Waugh, loc. cit.)

The previous example should be regarded as illustrative only. The observations are correlated in time, and it might be possible to obtain "nonsense" canonical correlations just as it is possible to obtain nonsense correlations in ordinary time-series analysis. The following example is free from this difficulty.

Information was available about 138 samples of Canadian Hard Red Spring wheat and the flour made from them. For the wheat five measurements were obtained:

X_1 kernel texture

X_2 test weight

X_3 damaged kernels

X_4 foreign material

X_5 crude protein in the wheat.

For the flour there were four measurements:

X_6 wheat per barrel of flour

X_7 ash in flour

X_8 crude protein in flour

X_9 gluten quality index.

Here $n = 138$, $p = 4$, $q = 5$. The correlation matrix was

	1	2	3	4	5	6	7	8	9
1	1·0	0·754 09	−0·690 48	−0·445 78	0·691 73	−0·604 63	−0·478 81	0·779 78	−0·152 05
2		1·0	−0·712 35	−0·514 83	0·411 84	−0·722 36	−0·418 78	0·542 45	−0·102 36
3			1·0	0·323 26	−0·443 93	0·737 42	0·361 32	−0·546 24	0·172 24
4				1·0	−0·334 39	0·527 44	0·460 92	−0·392 66	−0·018 73
5					1·0	−0·383 10	−0·504 94	0·736 66	−0·148 48
6						1·0	0·250 56	−0·489 93	0·249 55
7							1·0	−0·433 61	−0·078 51
8								1·0	−0·162 76
9									1·0

The first canonical correlation is $\pm 0{\cdot}909\,388$, an unusually high value. The first two canonical variates are

$$\xi_1 = \text{constant}\,(0{\cdot}357\,71X_1 + 0{\cdot}295\,08X_2 - 0{\cdot}560\,95X_3$$
$$- 0{\cdot}447\,40X_4 + 0{\cdot}504\,49X_5)$$

$$\eta_1 = \text{constant}\,(-X_6 - 0{\cdot}537\,27X_7 + 0{\cdot}847\,73X_8 + 0{\cdot}045\,78X_9), \qquad (5.32)$$

where, as usual, the X's are expressed in standard measure. These are the (linear) index-numbers which give us the maximum correlation between the properties of the wheat and those of the flour.

Here, again, having carried out the analysis, we need to look very carefully at the results to see if they make sense. On the whole it appears that they do. For example, in ξ_1 the signs given to kernel texture, test weight and crude protein are positive, these being the variables for which high scores indicate greater value — and those for the detrimental qualities of damaged kernels and foreign material are negative. In η_1 the wheat per barrel and ash content are negative and the crude protein and gluten quality positive. For these signs the canonical correlation is positive.

We could, of course, get equivalent results by changing all the signs of one of the canonical variables and taking the canonical correlations as $-0{\cdot}909\,388$. Which method of presentation we choose depends on the individual circumstances. The "constants" in (5.32) are positive.

It is, unfortunately, necessary to add that Waugh, to whom the analysis is due, carried out similar analyses on US Hard Red Winter wheat, and found that although the canonical correlations were higher than for Canadian, the signs of the coefficients in the canonical variables were no longer satisfactory in all cases. Waugh suggests that the variables were dominated by X_5 and X_8, representing gluten content, which was inversely related to some of the other desirable characteristics.

In work of this character there may arise instabilities in the individual coefficients of a type which we shall meet in Chapter 7 on Regression. To decide the question we ought to work out the other canonical roots and see whether collinear effects are present.

Example 5.4

The difficulties of interpretation are such that not many examples of convincing applications of canonical correlation analysis appear in the literature. We may notice a few which illustrate its potentialities in different fields.

(1) McGee (1965)
     A study was made of the way in which people pronounce certain vowels. A sound spectrogram of each individual was broken up into 34 bands ranging from 100 to 3500 cycles per second. Canonical analysis was applied on two vowels, "ee" as in heed and "aw" as in hawed. Conclusions were of two kinds, one that individual differences on the utterances of the vowels were much the same; the other that frequencies below 1000 Hz and above 2300 Hz are mainly responsible for the invariance of the personal characteristics of voices.

(2) Glahn (1968)
     For 30 meteorological stations at 00.00 GMT in June, July and August over five years, the 500 millibar (barometric pressure) height was observed on one day and on the previous day; $p = 30$ and $n = 455$. The canonical correlations were studied to see how far the situation at one time-point could be used to predict that 24 hours later.

(3) Adelman *et al.* (1969)
     For 74 underdeveloped countries ($= n$) a set of nine "goal" variables, e.g. the rate of growth of real *per capita* Gross National Product and the level of modernization of techniques in agriculture, were compared with 14 "instrumental" variables such as extent of literacy and change in the degree of industrialization ($p = 9$, $q = 14$). Canonical correlation analysis was applied to consider which instruments were most highly associated with which goals.

5.7 The general sampling problems associated with canonical correlations are very severe, but, as for principal components, large-sample results are fairly simple and can be obtained without much difficulty. Starting (and omitting the suffix i) from

$$\Sigma \, l_\alpha l_\beta c_{\alpha\beta} = 1 \qquad (5.33)$$

$$\Sigma \, m_a m_b c_{ab} = 1 \qquad (5.34)$$

$$\Sigma \, l_\alpha m_a c_{\alpha a} = r \qquad (5.35)$$

we have, on differentiating,

$$2 \Sigma c_{\alpha\beta} l_\alpha dl_\beta + \Sigma l_\alpha l_\beta dc_{\alpha\beta} = 0 \qquad (5.36)$$

$$2 \Sigma c_{ab} m_a dl_b + \Sigma l_a l_b dc_{ab} = 0 \qquad (5.37)$$

$$dr = \Sigma l_\alpha m_a dc_{\alpha a} + \Sigma l_\alpha c_{\alpha a} dm_a + \Sigma m_a c_{\alpha a} dl_\alpha. \qquad (5.38)$$

Without loss of generality we may now suppose the variables transformed to canonical correlation components, so that (concentrating for some particular i, say $i = 1$) we have

$$2dl_1 + dc_{11} = 0 \qquad (5.39)$$

$$2dm_1 + dc_{p+1,p+1} = 0 \qquad (5.40)$$

$$dr_1 = dc_{1,p+1} + r_1 dl_1 + r_1 dm_1. \qquad (5.41)$$

Substituting from the first two of these equations in the third, we have

$$dr_1 = dc_{1,p+1} - \tfrac{1}{2} r_1 (dc_{11} + dc_{p+1,p+1}). \qquad (5.42)$$

Similar equations apply to any other suffix, say 2:

$$dr_2 = dc_{2,p+2} - \tfrac{1}{2} r_2 (dc_{22} + dc_{p+2,p+2}), \qquad (5.43)$$

Multiplying (5.42) and (5.43), taking expectations, and using the results of (2.40), we have

$$\mathrm{cov}\,(r_1, r_2) = 0 \qquad (5.44)$$

$$\mathrm{var}\,r_1 = \frac{1}{n}(1 - r_1^2)^2. \qquad (5.45)$$

This, as it happens, is the same result as for an ordinary product-moment correlation coefficient, but it must not be assumed that the *distribution* of r_1 is the same as that of the ordinary correlation.

6

Some distribution theory[(*)]

6.1 The bivariate normal distribution is usually written in the form

$$dF = \frac{1}{2\pi(1-\rho^2)^{\frac{1}{2}}\sigma_1\sigma_2}$$

$$\times \exp\left[\frac{-1}{2(1-\rho^2)}\left\{\left(\frac{x_1-\mu_1}{\sigma_1}\right)^2 - 2\rho\left(\frac{x_1-\mu_1}{\sigma_1}\right)\left(\frac{x_2-\mu_2}{\sigma_2}\right) + \left(\frac{x_2-\mu_2}{\sigma_2}\right)^2\right\}\right]$$

$$dx_1\,dx_2. \qquad (6.1)$$

We shall usually lose no generality by supposing the variables measured from their respective means, so that (6.1) can be condensed to

$$dF = \frac{1}{2\pi(1-\rho^2)^{\frac{1}{2}}} \exp\left[\frac{-1}{2(1-\rho^2)}\left\{\frac{x_1^2}{\sigma_1^2} - \frac{2\rho x_1 x_2}{\sigma_1\sigma_2} + \frac{x_2^2}{\sigma_2^2}\right\}\right]\frac{dx_1\,dx_2}{\sigma_1\,\sigma_2}. \qquad (6.2)$$

This form, however, tends to obscure the path to a generalization to p dimensions. Consider the covariance matrix, say $\mathbf{Y}$:

$$\mathbf{Y} = \begin{bmatrix} \sigma_1^2 & \sigma_1\sigma_2\rho \\ \sigma_1\sigma_2\rho & \sigma_2^2 \end{bmatrix} \qquad (6.3)$$

Its determinant is $\sigma_1^2\sigma_2^2(1-\rho^2)$, so that the inverse, $\mathbf{Y}^{-1}$, say $\boldsymbol{\alpha}$, is given by

$$\boldsymbol{\alpha} = \begin{bmatrix} \dfrac{1}{\sigma_1^2(1-\rho^2)} & -\dfrac{\rho}{\sigma_1\sigma_2(1-\rho^2)} \\ \dfrac{-\rho}{\sigma_1\sigma_2(1-\rho^2)} & \dfrac{1}{\sigma_2^2(1-\rho^2)} \end{bmatrix} \qquad (6.4)$$

We may then write (6.2) simply as

[(*)] This chapter is added for the benefit of the mathematical statistician. It requires a good deal more background in mathematics and theoretical statistics than the remainder of the book and can be omitted by those willing to take the results on trust.

$$dF = \frac{|\alpha|^{\frac{1}{2}}}{2\pi} \exp\left\{-\tfrac{1}{2} \sum_{i,j=1}^{2} \alpha_{ij} x_i x_j\right\} dx_1\, dx_2. \tag{6.5}$$

This is the form which lends itself most readily to generalization. In fact, for p variables measured from their means, we define the multivariate normal distribution as

$$dF = k \exp\left\{-\tfrac{1}{2} \sum_{i,j=1}^{p} \alpha_{ij} x_i x_j\right\} \prod_{i=1}^{p} dx_i \tag{6.6}$$

$$= k \exp\left(-\tfrac{1}{2} x' \alpha\, x\right) \Pi\, dx, \tag{6.7}$$

x being a column vector and α, for the moment, a $p \times p$ symmetric matrix conditioned only by the fact that $\sum \alpha_{ij} x_i x_j$ must be non-negative definite (for if it were not, the integral of dF over the course of the x's would not converge). We evaluate the constant k, and show that α is in fact the inverse of the dispersion matrix, and proceed to find the characteristic function.

6.2   As in the case of principal components, we can always find a unitary linear transformation to new variables y such that $\sum \alpha_{ij} x_i x_j$ becomes $\sum \lambda_i y_i^2$, where λ's are the eigenvalues of the matrix α. The Jacobian of the transformation is unity, so (6.6) becomes

$$dF = k \exp\left(-\tfrac{1}{2} \sum \lambda_i y_i^2\right) \Pi\, dy_i. \tag{6.8}$$

A further transformation $y_i \lambda_i^{\frac{1}{2}} = z_i$ reduces this to

$$dF = \frac{k}{(\lambda_1 \lambda_2 \ldots \lambda_p)^{\frac{1}{2}}} \exp\left(-\tfrac{1}{2} \sum z_i^2\right) \Pi\, dz_i. \tag{6.9}$$

Now the product of the eigenvalues is the determinant of α. The multiple integral of dF from z_i ranging from $-\infty$ to $+\infty$ is the product of p independent normal integrals, each of which is $(2\pi)^{\frac{1}{2}}$. Thus we find $k = |\alpha|^{\frac{1}{2}}/(2\pi)^{\frac{1}{2}p}$, and the multivariate normal distribution is

$$dF = \frac{|\alpha|^{\frac{1}{2}}}{(2\pi)^{\frac{1}{2}p}} \exp\left(-\tfrac{1}{2} \sum \alpha_{ij} x_i x_j\right) \Pi\, dx_i. \tag{6.10}$$

6.3   Consider now

$$\frac{\partial}{\partial \alpha_{ij}} \int dF,$$

where the integral sign denotes the multiple integral over p variables. We have

$$0 = \frac{\partial}{\partial \alpha_{ij}} \int dF = \int -\tfrac{1}{2} x_i x_j\, dF + \int \frac{\partial |\alpha|}{2 \partial \alpha_{ij}} \frac{1}{|\alpha|} dF,$$

or the covariance of $x_i x_j$ is given by

$$\text{cov}(x_i x_j) = \frac{\partial |\alpha|}{|\alpha| \, \partial \alpha_{ij}} = \frac{A_{ij}}{|\alpha|}, \tag{6.11}$$

where A_{ij} is the co-factor of α_{ij} in α. Thus the expression on the right is the (i, j)th term in α^{-1}, and we have for the covariance matrix

$$\gamma = \alpha^{-1}. \tag{6.12}$$

6.4   For the bivariate normal distribution, measured from the means, it is known that the characteristic function is given by

$$\phi(t_1, t_2) = \exp \left\{ \tfrac{1}{2} \sum_{i,j=1}^{2} \gamma_{ij} t_i t_j \right\}. \tag{6.13}$$

This also is a form which generalizes to p dimensions, in the form

$$\phi(t_1, t_2, \ldots, t_p) = \exp \left\{ \tfrac{1}{2} \sum_{i,j=1}^{p} \gamma_{ij} t_i t_j \right\} \tag{6.14}$$

$$= \exp \{-\tfrac{1}{2} t' \gamma t\}. \tag{6.15}$$

In fact, if L is the matrix of the orthogonal transformation $x = Ly$, and Λ is the diagonal matrix of eigenvalues, we know that (as in equation (2.10))

$$\alpha L = L \Lambda$$

so that

$$L^{-1} \alpha^{-1} = \Lambda^{-1} L^{-1}$$

$$\alpha^{-1} = L \Lambda^{-1} L^{-1} = L \Lambda^{-1} L'. \tag{6.16}$$

The characteristic function is proportional to

$$\int \exp \left\{ -\tfrac{1}{2} \sum \alpha_{ij} x_i x_j + \sum it_j x_j \right\} \Pi \, dx_j$$

$$= \int \Pi \, dy_j \exp -\tfrac{1}{2} \left\{ \sum \lambda_j y_j^2 - 2 \sum_k it_k \sum_j l_{kj} y_j \right\}$$

reducing, on completing the squares of the individual y's, to

$$\left. \begin{array}{l} \exp \left[\tfrac{1}{2} \sum_j \left(\dfrac{\sum_k it_k \sum l_{kj}}{\lambda_i} \right)^2 \right] = \exp \left(-\tfrac{1}{2} t' L \Lambda^{-1} L' t \right) \\[2ex] \qquad = \exp \left(-\tfrac{1}{2} t' \alpha^{-1} t \right) \\[1ex] \qquad = \exp \left(-\tfrac{1}{2} t' \gamma t \right). \end{array} \right\} \tag{6.17}$$

Since this is unity when $t = 0$, it is in fact the characteristic function.

6.5   The multivariate normal distribution forms the basis of most of what is known about exact sampling theory. Like the univariate form of which it is a generalization, it is rarely encountered in practice, but a number of parental distributions approximate to it. In particular, under a generalization of the Central Limit Theorem, the means of many statistics tend to multivariate normality as n, the sample number, increases.

It may be convenient at this point to indicate which of the results in univariate theory generalize to the multivariate domain.

Univariate	*Multivariate*
Normal	Multivariate normal
Variance	Covariance (dispersion) matrix
Normal distribution of a mean	Joint multivariate normal distribution of means
Chi-squared distribution (distribution of variance)	Wishart's distribution (distribution of covariances)
Student's t (t^2 is square of ratio of mean to variance)	Hotelling's T^2 (ratio of determinant of quadratic form in means to covariance determinant)
Fisher's z (ratio of two variances)	Ratio of two covariance-type determinants
Maximum likelihood estimation	Maximum likelihood estimation
Analysis of variance (ANOVA)	Multivariate analysis of variance (MANOVA)

There are, of course, regions of study in multivariate analysis which have no univariate counterpart: component and factor analysis, cluster analysis, multiple contingency tables, multidimensional scaling, and so forth.

6.6   We also remark without proof that two useful properties of univariate sampling hold for the multivariate form:

(1)   In normal variation the joint distribution of means is independent of the joint distribution of covariances. The converse is also true, namely that if the two distributions for any one sample size are independent, the parent distribution must be normal.

(2)   For the joint estimation of parameters the maximum likelihood method asymptotically minimizes the determinant of sampling variances and covariances (as compared with the univariate result that, under certain regularity conditions, ML estimates asymptotically have minimal variance).

6.7   The geometrically minded reader may be interested in a spatial representation of the covariance determinant. In the Second Space referred to in **1.10** the square of the length of the ith vector through the origin is n times the second moment of x_i about zero; and if the origin is at the mean, the square of the length is n times the variance of x_i. The p vectors through the mean may be regarded as the edges of a hyperparallelogram (a parallelotope).

Then (we state without proof) the square of hypervolume (content) of this parallelotope is n^p times the determinant of covariances. Likewise, if the origin is at an arbitrary point, the parallelotope so defined has a volume-squared equal to n^p times the determinant of second-order moments about that point.

Many of the tests we shall consider in MANOVA can be regarded as the ratio of two determinants of the covariance type, and from the geometrical viewpoint tests of homogeneity or heterogeneity can be looked upon as relating to the extent to which two "contents" in the Second Space coincide or differ materially.

6.8 We shall state some of the basic results in sampling theory without proof (for which see Kendall and Stuart, vol. 3).

(1)   The joint distribution of means of samples of n from the population of equation (6.6) has the same form, except that the dispersion matrix is γ/n. In particular the correlation between $\bar{x}_i$ and $\bar{x}_j$ is the same as the correlation between x_i and x_j, namely ρ_{ij}.

(2)   The joint distribution of covariances, which is independent of the joint distribution of means, is given by Wishart's distribution:

$$dF = \frac{(\tfrac{1}{2}n)^{\frac{1}{2}p(n-1)}|\alpha|^{\frac{1}{2}(n-1)}|c|^{\frac{1}{2}(n-p-2)}}{\pi^{\frac{1}{4}p(p-1)} \prod\limits_{j=1}^{p} \Gamma\{\tfrac{1}{2}(n-j)\}} \exp\left(-\frac{n}{2}\sum\limits_{i,j=1}^{p}\alpha_{ij}c_{ij}\right) \prod\limits_{j\leqslant k}^{p} dc_{jk}.$$

(6.18)

(3)   The characteristic function of this distribution is

$$\phi(t_1, \text{etc.}) = \cfrac{|\alpha|^{\frac{1}{2}(n-1)}}{\begin{vmatrix} \alpha_{11}-2it_{11}/n & \alpha_{12}-it_{12}/n & \ldots & \alpha_{1p}-it_{1p}/n \\ \alpha_{21}-it_{21}/n & \alpha_{22}-2it_{22}/n & \ldots & \alpha_{2p}-it_{2p}/n \\ & & \ldots & \cdot \\ \alpha_{p1}-it_{p1}/n & \alpha_{p2}-it_{p2}/n & \ldots & \alpha_{pp}-2it_{pp}/n \end{vmatrix}^{\frac{1}{2}(n-1)}}$$

(6.19)

6.9 The Wishart distribution is not easy to handle. The limits of variation of the c_{ik}'s are not independent, so that we cannot, in general, integrate out a particular set in order to get the distribution of the remainder. However, from (6.19) expanded we can obtain as many moments as we please and hence fit a distribution if some particular c or combination of c's is of special interest. In practice this is rarely necessary.

6.10 We are sometimes concerned with the distribution of the dispersion determinant $|c|$ as a whole. Again, a closed explicit form is unobtainable, but we may derive its moments. In fact,

$$E \, |c|^t = \frac{2^{pt}}{n^{pt}} \prod_{j=1}^{p} \frac{\Gamma\{\frac{1}{2}(n-j)+t\}}{\Gamma\{\frac{1}{2}(n-j)\}} |\gamma|^t. \qquad (6.20)$$

6.11   The occurrence of products of Gamma-functions in expressions like (6.20) is common in this class of work and sometimes enables us to find approximative results based on the asymptotic expansion of the log Γ function. A particular case of some theoretical interest is the distribution of the correlation determinant $|r|$ *when all the parent population correlations are zero*, i.e. the variables in the parent distribution are independent. In such a case,

$$E \, |r|^t = \frac{[\Gamma\{\frac{1}{2}(n-1)\}]^p}{[\Gamma\{\frac{1}{2}(n-1)+t\}]^p} \frac{\prod\limits_{j=1}^{p} \Gamma\{\frac{1}{2}(n-j)+t\}}{\prod\limits_{j=1}^{p} \Gamma\{\frac{1}{2}(n-j)\}}. \qquad (6.21)$$

Rather surprisingly it turns out that $-(n-1) \log |r|$ is asymptotically distributed as χ^2 with $\frac{1}{2}p(p-1)$ degrees of freedom. We shall meet this type of χ^2 approximation on several subsequent occasions. It is gratifying to find that, notwithstanding the complexity of expressions like (6.18) and (6.19), practical tests can be carried out by reference to the χ^2 tables, and no other distributional calculations are required, at least for large samples. On the other hand, (6.21) is based on the assumption that the parent variables are independent, which is usually the case of least interest.

6.12   In univariate theory we sometimes test the significance of a mean by the statistic known as Student's t. Its usefulness in normal samples is based on the fact that its distribution is independent of the parent variance σ^2. It is in fact the ratio of two statistics of comparable scale and may be written as

$$t^2 = \frac{(n-1)\bar{x}^2}{s^2}, \qquad (6.22)$$

where s^2 is the sample variance, defined as $\sum\limits_{i=1}^{n} (x_1 - \bar{x})^2/n$. In generalization, putting

$$d_{jk} = nc_{jk}, \qquad (6.23)$$

we define Hotelling's T^2 as

$$T^2 = n(n-1) \sum_{j,k=1}^{p} D_{jk} \bar{x}_j \bar{x}_k, \qquad (6.24)$$

where D is the inverse of d. It may be shown that

$$\frac{(n-p)T^2}{p(n-1)} \quad \text{has an } F(p, n-p) \text{ distribution}, \qquad (6.25)$$

that is to say, may be tested in the F-distribution (the variance-ratio distribution) with p and $n-p$ degress of freedom.

The utility of the test resides in the fact that if we have a set of correlated x's, tests on the individual means of the x's are not independent and hence we cannot logically apply a set of t-tests. T^2 enables us to test a hypothesis concerning all the means at once. Thus, if the hypothesis is that x_i has mean $m_i, i = 1, 2, \ldots, p$, we test the statistic

$$n(n-1) \sum_{j,k=1}^{p} D_{jk}(\bar{x}_j - m_j)(\bar{x}_k - m_k) \qquad (6.26)$$

and accept or reject according to its significance values in the usual way.

6.13 It is natural to inquire whether there exists a multivariate generalization of the other test so often employed in univariate theory, the variance-ratio, F-ratio, or the equivalent Fisher z. This also, being the ratio of two statistics in the same scale, is independent of the parent variance in normal samples. There are indeed generalizations, but we defer a consideration of them until Chapter 9, where they arise naturally in the testing of hypotheses in MANOVA.

6.14 In view of the intractability of many multivariate sampling distributions, results for large samples are by no means to be despised. Let us derive the sampling variance of the second-order terms, variance and covariance of the sample.

To the first order in n we have

$$E(c_{jk}) = \gamma_{jk}, \qquad (6.27)$$

where c refers to the covariance of the sample and γ to the parent. To the same order, the product-moment of two covariance terms,

$$E(c_{jk}c_{lm}) = \frac{1}{n^2} E \left\{ \sum_{\alpha=1}^{n} x_{j\alpha} x_{k\alpha} \sum_{\beta=1}^{n} x_{l\beta} x_{m\beta} \right\}. \qquad (6.28)$$

If $\alpha \neq \beta$ the two sums on the right are independent. There are $n(n-1)$ terms in the resulting product, and the expectation of each is $\gamma_{jk}\gamma_{lm}$. If $\alpha = \beta$ we have to consider n terms such as $E(x_{j\alpha} x_{k\alpha} x_{l\alpha} x_{m\alpha})$. In general this involves moments of the fourth order. We therefore specialize to normal variation. From the characteristic function (6.14) we have, on expansion, from the coefficient of $t_j t_k t_l t_m$,

$$E(x_{j\alpha} x_{k\alpha} x_{l\alpha} x_{m\alpha}) = \gamma_{jk}\gamma_{lm} + \gamma_{jm}\gamma_{kl} + \gamma_{jl}\gamma_{km}. \qquad (6.29)$$

Then, from (6.28),

$$E(c_{jk}c_{lm}) = \gamma_{jk}\gamma_{lm} + \frac{1}{n} \{\gamma_{jm}\gamma_{kl} + \gamma_{jl}\gamma_{km}\}. \qquad (6.30)$$

Subtracting $E(c_{jk}) E(c_{lm})$, we find

$$\text{cov}\,(c_{jk}, c_{lm}) = \frac{1}{n}(\gamma_{jm}\gamma_{kl} + \gamma_{jl}\gamma_{km}). \tag{6.31}$$

In particular, if $j = k = l = m$ we have

$$\text{var}\,c_{jj} = \frac{2\gamma_{jj}^2}{n}, \tag{6.32}$$

and if $j = l$, $k = m$,

$$\text{var}\,c_{jk} = \frac{1}{n}(\gamma_{jk}^2 + \gamma_{jj}\gamma_{kk}). \tag{6.33}$$

In using these formulae, of course, we insert the sample values c for the γ's on the right. They are valid only for normal distributions.

6.15 In testing individual parameters one by one, however, we must always remember, as remarked in **6.12**, that in general the tests are correlated in virtue of correlation in the parent. More often, perhaps, in multivariate analysis, we are concerned to test a hypothesis concerning structure as a whole rather than an individual constituent of it: for example, whether one set of variables is independent of the remainder or whether certain components are "significant". Some of these tests will be dealt with in Chapter 9. For the remainder of this chapter we are concerned with the "significance" of principal components and factors.

6.16 The actual distribution of eigenvalues of a covariance matrix is exceedingly complicated, even when all parent correlations vanish, and is unknown in the more general case. One reason is that the eigenvalues are not rational symmetric functions of the covariances. They are the roots of a p-ic equation and in consequence are transcendental. As we have remarked, certain functions of the roots, for example their sum (the trace of the covariance matrix) and their product (the determinant of the matrix), are more tractable than the roots themselves and are sometimes used in tests of structure.

6.17 Another unusual feature concerns the meaning which we attribute to the word "significant" in speaking of a non-zero eigenvalue. Strictly speaking, any such value, however small, is significant in the sense that it could not have arisen from a population in which the corresponding parent value is zero. For example, if the parent $\lambda_p = 0$ the scatter of points lies in a hyperplane (collapses into one lower dimension) and no sample can do otherwise. It is true that some errors of observation superposed on the variate values in the sample may create the illusion of an extra dimension; but we have been supposing that our variables are not subject to observational error. (Something about the contrary case will be said in Chapter 8 on problems of curve fitting.)

6.18 We may, however, ask a rather different question: do the roots differ significantly among themselves? If they do not, then they could have arisen from a parent in which all the eigenvalues are equal, in which case the variables are uncorrelated, no transformation to principal components is necessary, and one orthogonal transformation is as good as any other. Or, as an extension of the same point, if we find that the first k components are unequal, is there any evidence that the remaining $p - k$ differ among themselves? This approach was used by Bartlett (1954 and earlier papers) to suggest a somewhat heuristic test of the distinguishability of principal components through the eigenvalues.

6.19 Suppose that we have, in fact, decided that the k largest eigenvalues are different. Assuming that the sample is large enough for us to be able to associate the sample λ's with their corresponding parent λ's, it seems reasonable to test the correlation determinant (which is the product of the eigenvalues) against the determinant which would have been reached had the last $p - k$ values been equal. Now the sum of the eigenvalues is p in a correlation determinant, so the average of the last $p - k$ is

$$\frac{p - \sum_{i=1}^{k} \lambda_i}{p - k}.$$

The correlation determinant with its last $p - k$ eigenvalues of this amount is then

$$\prod_{i=1}^{k} \lambda_i \left\{ \frac{p - \sum_{i=1}^{k} \lambda_i}{p - k} \right\}^{p-k}. \tag{6.34}$$

The ratio of this to the observed determinant $\prod_{i=1}^{p} \lambda_i$ is then

$$\left(\frac{\sum_{i=k+1}^{p} \lambda_i}{p - k} \right)^{p-k} \Big/ \prod_{i=k+1}^{p} \lambda_i. \tag{6.35}$$

This may be regarded as the $(p - k)$th power of the ratio of arithmetic to geometric means of the smallest $(p - k)$ eigenvalues. It may be shown that $(n - k)$ times the logarithm of this quantity may be tested in the χ^2 distribution with $\frac{1}{2}(p - k - 1)(p - k + 2)$ degrees of freedom. Actually this number of d.f. is appropriate when the covariance matrix is being tested, not the correlation matrix. Bartlett (1951b) showed that in the latter case the number of d.f., even asymptotically, depends on the amount of variance abstracted in the first k components; and further, that the number $\frac{1}{2}(p - k - 1)(p - k + 2)$ is an upper limit to the d.f., so that a test based on it is conservative. We note in particular that when $k = 0$ the number of d.f. is $\frac{1}{2}(p-1)(p+2)$, whence that for a correlation matrix is $\frac{1}{2}p(p-1)$ – see **6.11**.

Example 6.1 (data from Lawley and Maxwell, 1971)

Five psychological tests were carried out on 123 individuals. The correlation matrix of scores was

		Test		
1	2	3	4	5
1·000	0·438	−0·137	0·205	−0·178
	1·000	0·031	0·180	−0·304
		1·000	0·161	0·372
			1·000	−0·013
				1·000

$$(6.36)$$

The eigenvalues were

$$1·757\,14, \quad 1·330\,70, \quad 0·780\,86, \quad 0·709\,16, \quad 0·422\,14.$$

The determinant of the correlation matrix is 0·546 59.

We might first inquire whether (notwithstanding their numerical differences) such a set of eigenvalues could arise from a parent in which they were all equal. For this purpose **(6.11)**, $(n-1) \log |r|$ is tested in χ^2 with $\frac{1}{2}p(p-1) = 10$ d.f. The actual value of the test statistic is $-122 \log (0·546\,59) = 73·7$ and is highly significant. We conclude that one – at least the largest – of the eigenvalues is significantly different from the rest. We might then go on and test with (6.35) and $k = 1$ to see whether the other four are significantly different. We should decide that the second eigenvalue is so. For the last three roots (product 0·233 76, mean 0·637 39) the criterion becomes 120 {3 log 0·637 39 − log 0·233 76} = 12·3, with five degrees of freedom. This exceeds the five percent but not the one percent significance value. We suspect that the last three roots are unequal, but the evidence is not overwhelmingly strong.

6.20 There are a number of minor refinements in the literature of the formulae we have just exemplified. It is doubtful whether they are worth making, but to avoid confusion we quote two of them.

(1)   Bartlett (1954)

$$-n \left(1 - \frac{2p + 11}{6n}\right) \log |r| \text{ is distributed as } \chi^2 \text{ with } \tfrac{1}{2}p(p-1)\,\text{d.f.} \quad (6.37)$$

(2)   Lawley (1956)

The criterion (6.35) multiplied by

$$n - k - 1 - \frac{1}{6}\left\{\frac{2(p-k)^2 + p - k + 2}{p - k}\right\} + \lambda^2 \sum_{j=1}^{k} \frac{1}{(\lambda_j - \lambda)^2} \qquad (6.38)$$

has the correct moments of a χ^2 distribution as far as and including the order n^{-3}. Here λ is the mean of $\lambda_{k+1}, \ldots, \lambda_p$.

6.21 The above results apply to the eigenvalues of a *correlation* matrix. Those of a covariance matrix are more complicated and indeed it seems that analogous statistics for a covariance matrix do not follow a χ^2 distribution. In the present state of knowledge there seems little alternative to working on the correlation matrix, notwithstanding what was said in Chapter 3 about the effect of scale on the eigenvalues of a dispersion matrix.

6.22 Let us now consider the analogous problem of factor analysis. We have here, in fact, two problems, first to estimate the l's and Σ, and second to decide how many factors; or, perhaps, in the reverse order.

We begin with the likelihood function of the covariances, distributed in the Wishart form of (6.18). The part dependent on the parameters may be written

$$\log L = -\tfrac{1}{2}(n - 1) \log |\gamma| - \tfrac{1}{2}n \sum_{j,k=1}^{p} \alpha_{jk} c_{jk}. \qquad (6.39)$$

As usual, we define a covariance as typified by

$$c_{jk} = \frac{1}{n} \Sigma \, (x_j - \bar{x}_j)(x_k - \bar{x}_k).$$

Here it is more convenient to define it with $n - 1$ instead of n in the denominator, in which case

$$\log L = \tfrac{1}{2}(n - 1)\{-\log |\gamma| - \Sigma \, \alpha_{jk} c_{jk}\}. \qquad (6.40)$$

Now we know from equation (4.8) that for the parent covariance matrix Y, inverse to α

$$Y = ll' + \Sigma. \qquad (6.41)$$

To estimate l and Σ we substitute in (6.40) for α and equate the differential coefficients to zero. We shall pass over the mathematical details rather briefly because the resulting equations, though interesting, do not seem to lend themselves very readily to numerical computation.

Differentiation with respect to σ_j^2 leads, after some considerable algebra, to

$$\hat{\alpha}_{ii} - \sum_{j,k} \hat{\alpha}_{ij} c_{jk} \hat{\alpha}_{ki} = 0, \qquad (6.42)$$

where we write carets to denote ML estimators. This is equivalent to

$$\mathrm{diag}\,(\hat{Y}^{-1} - \hat{Y}^{-1} c \, \hat{Y}^{-1}) = 0. \qquad (6.43)$$

Differentiation with respect to l_{jk} likewise gives

$$\sum_i \hat{l}_{ik} \hat{\alpha}_{jk} - \sum_{i,u,v} \hat{l}_{ik} \hat{\alpha}_{iu} c_{uv} \hat{\alpha}_{vj} = 0, \tag{6.44}$$

which is the element of the jth row and kth column of the $m \times p$ matrix

$$\hat{l}' \hat{\gamma}^{-1} - \hat{l}' \hat{\gamma}^{-1} c \hat{\gamma}^{-1} = 0. \tag{6.45}$$

Postmultiplication by $\hat{\gamma}$ yields

$$\hat{l}' - \hat{l}' \hat{\gamma}^{-1} c = 0 \tag{6.46}$$

and premultiplication by $\hat{l}$ gives the $p \times p$ matrix

$$\hat{l}\hat{l}' - (\hat{l}\hat{l}') \hat{\gamma}^{-1} c = 0. \tag{6.47}$$

It now appears necessary to assume that the ML estimators obey the same relation as the parent parameter, namely that

$$\hat{\gamma} = \hat{l}\hat{l}' + \hat{\Sigma}. \tag{6.48}$$

Premultiplying the diagonal matrix $\hat{\Sigma}$ $(= \hat{\gamma} - \hat{l}\hat{l}')$ into (6.43), we have, on reduction, in virtue of (6.47),

$$\text{diag}\,(\mathbf{I} - c\hat{\gamma}^{-1}) = 0$$

and postmultiplying again by $\hat{\gamma}$ we find that

$$\text{diag}\,(c - \hat{\gamma}) = 0. \tag{6.49}$$

This is equivalent to

$$\hat{\sigma}_j^2 = c_{jj} - \sum_{k=1}^{m} \hat{l}_{jk}^2 \tag{6.50}$$

and will provide estimators of the σ^2 when we have those of the l's.

From (6.47) and (6.48) we find

$$\hat{l}\hat{l}' - (\hat{\gamma} - \hat{\Sigma}) \hat{\gamma}^{-1} c = 0.$$

Premultiply by $\hat{l}' \hat{\Sigma}^{-1}$ to obtain

$$\hat{l}' \hat{\Sigma}^{-1} \hat{l}\hat{l}' + \hat{l}' \hat{\gamma}^{-1} c - \hat{l}' \hat{\Sigma}^{-1} c = 0$$

or from (6.46)

$$\hat{l}' \hat{\Sigma}^{-1} \hat{l}\hat{l}' - \hat{l}' \hat{\gamma}^{-1} c + \hat{l}' = 0$$

$$(\hat{l}' \hat{\Sigma}^{-1} \hat{l} + \mathbf{I}) \hat{l}' - \hat{l}' \hat{\Sigma}^{-1} c = 0. \tag{6.51}$$

We now make an assumption similar, but not identical, to that of equation (4.16), namely that $\hat{l}' \hat{\Sigma}^{-1} \hat{l}$ is a diagonal matrix, say $\hat{\mathbf{J}}$; then

$$\hat{\mathbf{J}} = \hat{l}' \hat{\Sigma}^{-1} \hat{l} \tag{6.52}$$

$$(\mathbf{I} + \hat{\mathbf{J}}) \hat{l}' = \hat{l} \hat{\Sigma}^{-1} c$$

and, taking transposes,

$$\hat{l}(\mathbf{I} + \hat{\mathbf{J}}) = c \hat{\Sigma}^{-1} \hat{l} \tag{6.53}$$

$$\hat{l}\hat{J} = c\hat{\Sigma}^{-1}\hat{l} - \hat{l}$$
$$= (c - \hat{\Sigma})\hat{\Sigma}^{-1}\hat{l}. \tag{6.54}$$

6.23 Recalling that $\hat{J}$ is diagonal, we see that its elements are the eigenvalues of $(c - \hat{\Sigma})\hat{\Sigma}^{-1}$. One would then suppose that the solution might proceed by iteration. If we can guess $\hat{\Sigma}$, equation (6.54) gives us $\hat{l}$. This can be used from (6.50) to make a second estimate of $\hat{\Sigma}$; and so on.

For reasons which have not been fully explained, such a procedure seems, not infrequently, either to take an unacceptable number of iterations or to converge on zero values of some of the Σ's. Jöreskog (1969) has tackled the problem afresh by directly (again by iteratively) assaulting the likelihood function. His program is reported to work well and we used it in Chapter 4 to illustrate some practical data.

6.24 The likelihood approach also gives us a test of the adequacy of m, the number of factors. The criterion is derivable by the methods we shall develop in Chapter 9 for testing hypotheses. We quote it without proof. The quantity defined as

$$\chi^2 = -n\left\{\log\frac{|c|}{|\hat{\gamma}|} - \text{trace}(c\gamma^{-1}) + p\right\} \tag{6.55}$$

is approximately distributed as the ordinary χ^2 with

$$\tfrac{1}{2}\{(p - m)^2 - (p + m)\} \quad \text{degrees of freedom.} \tag{6.56}$$

If for a given m the value of χ^2 is significantly large, more factors are required. We note from equation (4.13) that the number of degrees of freedom will not be negative.

Bartlett (1951b) suggested that a better approximation would be obtained by using, instead of n in (6.55), the multiplier

$$n' = n - \tfrac{1}{6}(2p + 11) - \tfrac{2}{3}m. \tag{6.57}$$

Lawley and Maxwell (1971) propose a simpler form than (6.55):

$$n' \sum_{j<k} \frac{(c_{jk} - \hat{\gamma}_{jk})^2}{(\hat{\sigma}_j\hat{\sigma}_k)^2}. \tag{6.58}$$

The Jöreskog program referred to carries out the test automatically.

NOTES

(1) The assumption that J is diagonal is for computational convenience only and resolves the indeterminacy in the factors in the same arbitrary way which we used in Chapter 3. There may therefore be a case for rotating the estimates so found by one of the methods mentioned in that chapter.

(2)   It is, in fact, possible to go further and impose a pattern on the l-matrix
      at the outset, e.g. by requiring that certain variables do appear in cer-
      tain factors or that others do not. Although the subject has not been
      fully explored, it appears that solutions can be found by maximizing
      the likelihood under such constraints (Jöreskog, 1970). The imposition
      of prior structure, of course, runs somewhat counter to the exploratory
      nature of the type of factor analysis which sets out without restriction.

(3)   Care must be taken with software written for computers when carrying
      out the type of analysis considered in this chapter. Some of the con-
      clusions seem to be sensitive to small errors in the estimates of para-
      meters.

(4)   For the further mathematics of distributional theory see T.W. Anderson
      (1958) and Kendall and Stuart, vol. 3.

7

Problems in regression analysis

7.1   The reader is assumed to be familiar with the elementary theory of regression analysis, the basic results of which we briefly rehearse. The model is

$$y = \sum_{j=0}^{p} \beta_j x_j + \epsilon, \qquad (7.1)$$

where the β's are under estimate and ϵ is a random variable independent of the regressors x with zero mean. x_0 may be regarded as a dummy variable equal to unity. It is often convenient to measure the regressand y and the regressors x about their means, in which case β_0 is zero. If there are n observations, we may represent the data in matrix form as

$$\underset{1 \times n}{y} = \underset{1 \times p}{\beta} \underset{p \times n}{x} + \underset{1 \times n}{\epsilon}. \qquad (7.2)$$

The constants β are estimated by least squares to give estimators b:

$$b = yx'(xx')^{-1}. \qquad (7.3)$$

The b are unbiased estimators, namely

$$E(b) = \beta \qquad (7.4)$$

and they have minimal variance in the class of estimators which are linear in y, their covariance matrix being given by

$$V(b) = (xx')^{-1}\sigma^2, \qquad (7.5)$$

where σ^2 is the variance of ϵ.

The sum of squares of fitted residuals is given by

$$\Sigma (y - \Sigma b_j x_j)^2 = \Sigma y^2 - \sum_j b_j \Sigma yx_j. \qquad (7.6)$$

The "goodness of fit" of the regression line is conveniently summarized in the ratio $\sigma^2/\text{var } y$ or equivalently in the square of the multiple correlation coefficient R^2:

$$R^2 = 1 - \frac{\sigma^2}{\text{var } y}. \qquad (7.7)$$

85

7.2   Two practical points need emphasis:

(1)   The estimators **b** of (7.3) depend on the *inverse* of the covariance matrix of the x's. If therefore the covariance matrix is degenerate, the estimators are indeterminate; and if the covariance determinant is small the estimators are very unreliable.

(2)   The sum of squares of residuals should be divided by $n - p$ to give an unbiassed estimator of σ^2. Machine programs sometimes give a value of R^2 taking into account the degrees of freedom, e.g.

$$R^2(\text{est.}) = 1 - \frac{\Sigma\,(\text{residuals squared})}{n-p} \cdot \frac{n-1}{\Sigma\,(y-\bar{y})^2} \qquad (7.8)$$

and sometimes what is described as a maximum-likelihood estimator:

$$R^2(\text{ML}) = 1 - \frac{\Sigma\,(\text{residuals squared})}{\Sigma\,(y-\bar{y})^2}. \qquad (7.9)$$

The two may differ substantially if R^2 is small and the residual element large compared with y.

7.3   In a study of various methods of computing regression coefficients Longley (1967) pointed out that algorithms in use on desk calculators are not necessarily the best for electronic machines. A later study by Wampler (1969) extended the inquiry. Longley ran some test data through a number of software programs which were available at the time and obtained quite different results from different programs. The trouble seems to stem from rounding-off errors in the inversion of the matrix $\mathbf{xx'}$. The statistician is not expected to be an expert on software programs, but it is desirable that he should know how they work in order to be satisfied about their reliability.

Measuring from the means, writing x_{p+1} for y, and writing d_{ij} for the sum of squares and cross-products about means (n times the covariances), we have for the estimating equations

$$0 = d_{11}b_1 + d_{12}b_2 + \ldots + d_{1p}b_p - d_{1,p+1} \qquad (7.10)$$

$$0 = d_{21}b_1 + d_{22}b_2 + \ldots + d_{2p}b_p - d_{2,p+1}, \qquad (7.11)$$

$$\text{etc.,}$$

to which we may adjoin

$$S = d_{p+1,1}\,b_1 + d_{p+1,2}\,b_2 + \ldots + d_{p+1,p}\,b_p - d_{p+1,p+1}. \qquad (7.12)$$

For reasons which will become apparent it is convenient to consider a more general set

$$v_1 = d_{11}u_1 + d_{12}u_2 + \ldots + d_{1p}u_p + d_{1,p+1}u_{p+1} \tag{7.13}$$

$$v_2 = d_{21}u_1 + d_{22}u_2 + \ldots + d_{2p}u_p + d_{2,p+1}u_{p+1} \tag{7.14}$$

. . . . . .

$$v_{p+1} = d_{p+1,1}u_1 + d_{p+1,2}u_2 + \ldots + d_{p+1,p}u_p + d_{p,p+1}u_{p+1}. \tag{7.15}$$

We first express u_1 in terms of v_1 and the remaining u's from (7.13) to obtain

$$u_1 = \frac{v_1}{d_{11}} - \frac{d_{12}}{d_{11}}u_2 - \ldots - \frac{d_{1,p+1}}{d_{11}}u_{p+1}. \tag{7.16}$$

(If d_{11} were zero we should rearrange the equations to bring one non-vanishing coefficient of u_1 to the top.) Substituting from (7.16) in the other · equations, we get a set

$$v_2 = \frac{d_{21}}{d_{11}}v_1 + \left(d_{22} - \frac{d_{21}d_{12}}{d_{11}}\right)u_2 + \ldots + \left(d_{2,p+1} - \frac{d_{21}d_{1,p+1}}{d_{11}}\right)u_{p+1},$$

$$\text{etc.,} \tag{7.17}$$

or, writing

$$d'_{k1} = d_{k1}/d_{11}, \quad d'_{1k} = -d_{1k}/d_{11}$$

$$d'_{k1} = d_{k1}/d_{11}, \quad d'_{jk} = d_{jk} - d_{j1}d_{1k}/d_{11}, \quad j,k \neq 1, \tag{7.18}$$

we have

$$u_1 = d'_{11}v_1 + d'_{12}u_2 + \ldots + d'_{1,p+1}u_{p+1} \tag{7.19}$$

$$v_2 = d'_{21}v_1 + d'_{22}u_2 + \ldots + d'_{2,p+1}u_{p+1} \tag{7.20}$$

. . . . . .

$$v_{p+1} = d'_{p+1,1}v_1 + d'_{p+1,2}u_2 + \ldots + d'_{p+1,p+1}u_{p+1}. \tag{7.21}$$

Whereas in equations (7.13)–(7.15) the formulae are symmetric in the sense that $d_{jk} = d_{kj}$, in (7.19)–(7.21) they are not. We can, however, restore the symmetry by adjoining a minus sign to v_1 so that the equations become

$$u_1 = \bar{d}_{11}(-v_1) + \bar{d}_{12}(u_2) + \ldots + \bar{d}_{1,p+1}u_{p+1} \tag{7.22}$$

$$v_2 = \bar{d}_{21}(-v_1) + \bar{d}_{22}(u_2) + \ldots + \bar{d}_{2,p+1}u_{p+1}, \tag{7.23}$$

$$\text{etc.,}$$

where

$$\left.\begin{array}{l} \bar{d}_{11} = -1/d_{11}, \quad \bar{d}_{1k} = \bar{d}_{k1} = -d_{1k}/d_{11} \\ \bar{d}_{jk} = \bar{d}_{kj} = d_{jk} - d_{j1}d_{1k}/d_{11}. \end{array}\right\} \tag{7.24}$$

The process of transforming the equations in this way is said to be one of "pivoting" on the first diagonal element d_{11}.

We can now repeat the process by pivoting on d_{22}, and so on, the coefficients being transformed after the manner of (7.24). Continuing the process and writing $\bar{d}$ for the coefficients so derived, we find finally

$$u_1 = \bar{d}_{11}(-v_1) + \bar{d}_{12}(-v_2) + \ldots + \bar{d}_{1p}(-v_p) + \bar{d}_{1,p+1} u_{p+1} \qquad (7.25)$$

$$u_2 = \bar{d}_{21}(-v_1) + \bar{d}_{22}(-v_2) + \ldots + \bar{d}_{2p}(-v_p) + \bar{d}_{2,p+1} u_{p+1} \qquad (7.26)$$

$$u_p = \bar{d}_{p1}(-v_1) + \bar{d}_{p2}(-v_2) + \ldots + \bar{d}_{pp}(-v_p) + \bar{d}_{p,p+1} u_{p+1}, \qquad (7.27)$$

with a further equation for v_{p+1}. Putting all the v's equal to zero and $u_{p+1} = -1$, we see from (7.10) that the first p of these equations simply yield the normal estimating equations, so that

$$b_j = -\bar{d}_{j,p+1}, \quad j = 1, 2, \ldots, p. \qquad (7.28)$$

Also the last equation gives us

$$S = -d_{p+1,p+1}, \qquad (7.29)$$

giving the residual sum of squares.

7.4   Two advantages of this systematic procedure are as follows. Suppose, first, that we stop after k operations, obtaining

$$u_1 = \bar{d}_{11}(-v_1) + \ldots + \bar{d}_{1k}(-v_k) + \bar{d}_{1,k+1} u_{k+1} + \ldots + \bar{d}_{1,p+1} u_{p+1}$$
$$(7.30)$$

$$u_k = \bar{d}_{k,1}(-v_1) + \ldots + d_{kk}(-v_k) + \bar{d}_{k,k+1} u_{k+1} + \ldots + \bar{d}_{k,p+1} u_{p+1}$$
$$(7.31)$$

$$v_{k+1} = \bar{d}_{k+1,1}(-v_1) + \ldots + \bar{d}_{k+1,k}(-v_k) + \bar{d}_{k+1,k+1} u_{k+1} + \ldots$$
$$+ \bar{d}_{k+1,p+1} u_{p+1} \qquad (7.32)$$

$$v_{p+1} = \bar{d}_{p+1,1}(-v_1) + \ldots + \bar{d}_{p+1,k}(-v_k) + \bar{d}_{p+1,k+1} u_{k+1} + \ldots$$
$$+ \bar{d}_{p+1,p+1} u_{p+1}. \qquad (7.33)$$

If we are interested in retaining in the regression only x_1 to x_k, β_{k+1} to β_p are zero. Putting $v_1 = v_2 \ldots = v_k = 0$ and $u_{k+1} = u_{k+2} = \ldots = 0$ in (7.30) to (7.32) we estimate b_1 to b_k.

7.5   Secondly, the diagonal elements $\bar{d}_{jj}$ for $j > k$ have an important interpretation. In fact, $-\bar{d}_{jj}$ is the residual sum of squares if x_j is (in the role of y) regressed on x_1 to x_k. The size of the diagonal coefficients then gives us a measure of the closeness of x_j to being a linear function of x_1 to x_k. In arithmetical practice a small value of d_{jj} below some tolerance level is a warning of possible degeneracy and indicates that we should be wasting our time in estimating further b's.

7.6   The order in which we pivot is at choice (or, to put it another way, we can please ourselves which of the labels 1 to p we attach to which x's).

It is natural to choose the largest coefficient of v_j when pivoting on that variable. The order of introduction of variables, however, is a separate subject which we discuss later in the chapter.

7.7 The sum of squares of residuals gives us some idea of the goodness of fit, as already remarked, and in pre-computer days was often the only criterion applied for that purpose. With computer aid, however, we can nowadays work out the n individual residuals, and it is always worth while to do so.

(1) The procedure will exhibit unusual "maverick" observations which may be playing a heavy role in the residual sum of squares. We then have to refer, if possible, to the provenance of such observations to see if some error has occurred in measurement or recording. In extreme cases we may reject the observation as atypical.

(2) Runs of positive or negative signs in the residuals are an indication that something in the model is inadequate.

(3) A frequency distribution of residuals gives us some idea whether the variable is skew.

7.8 It must be remembered that the observed residuals are not the same as the real (unobserved) residuals, and can only be regarded as giving a general indication of the behaviour of the latter. In fact, the observed residuals are a kind of weighted average of the ϵ's.

If we write **e** for the observed residuals,

$$\mathbf{y} = \mathbf{bx} + \mathbf{e} \tag{7.34}$$

as against

$$\mathbf{y} = \mathbf{Bx} + \boldsymbol{\epsilon}.$$

Hence

$$\boldsymbol{\epsilon} - \mathbf{e} = (\mathbf{b} - \mathbf{B})\mathbf{x}.$$

But

$$\mathbf{b} = \mathbf{yx'}(\mathbf{xx'})^{-1} = (\mathbf{Bx} + \boldsymbol{\epsilon})\mathbf{x'}(\mathbf{xx'})^{-1}$$

$$= \mathbf{B} + \boldsymbol{\epsilon}\,\mathbf{x'}(\mathbf{xx'})^{-1}.$$

Hence

$$\boldsymbol{\epsilon} - \mathbf{e} = \boldsymbol{\epsilon}\,\mathbf{x'}(\mathbf{xx'})^{-1}\mathbf{x}$$

$$\mathbf{e} = \boldsymbol{\epsilon}\,\{\mathbf{I}_n - \mathbf{x'}(\mathbf{xx'})^{-1}\mathbf{x}\}. \tag{7.35}$$

e, the matrix on the right in (7.35), is of rank $n - p$, corresponding to the fact that the **e**'s are distributed as χ^2 with $n - p$ degrees of freedom. In fact, $\mathbf{x}\{\mathbf{I}_n - \mathbf{x'}(\mathbf{xx'})^{-1}\mathbf{x}\} = 0$, giving p relations connecting the elements of the matrix.

7.9 In particular, if our observations are given in some sort of order (e.g. temporally) and we wish to test the hypothesis that $\boldsymbol{\epsilon}$ is a random variable, we cannot test the correlations between successive values of **e** against the

hypothesis that they are zero, because the averaging operation implicit in
(7.35) generates correlations among the e's even where they do not exist
among the ε's. This difficulty has been partly removed by a test due to
Durbin and Watson (1971 and previous papers). The test statistic is

$$\frac{\Sigma (e_{i+1} - e_i)^2}{\Sigma e_i^2}. \tag{7.36}$$

The actual distribution is complicated, but Durbin and Watson were able to
show that it lies between two known distributions. For details, reference
may be made to their paper and to a summary in Kendall (1973).

7.10 Suppose now that, from general scrutiny or more sophisticated procedures we are led to doubt the adequacy of the model. There arise several
points for consideration.

(1) R^2 is too small to give a good fit. In such a case, the obvious recourse
is to add more regressors.

(2) The relationship is not linear. We then have to consider whether to
import among the regressors terms of a nonlinear kind such as x_1^2 or
$x_1 x_2$, and possibly terms of higher degree; or, alternatively, whether
the regression can be brought to linearity by a variate transformation,
e.g. by taking logarithms. Usually a certain amount of trial and error
is necessary in such cases.

(3) The residuals are not random. This can be a very refractory case. The
indication is that we have omitted from the regressors some systematic
effect which has become incorporated into the residuals. Sometimes
this leads us to look for further regressors. Sometimes we have to amend
the model so as to allow for the behaviour of the residuals. Particularly,
when the observations are arranged in temporal order we may wish to
postulate connectivity in the residuals by some such relation as the
Markoff scheme

$$\epsilon_{t+1} = \rho \epsilon_t + \eta_t, \tag{7.37}$$

where ϵ_t is the residual at time t and η_t is itself a random term. The
analysis of such series is better handled as a problem in time-series and
reference may be made to Kendall (1973) for some of the possible
modes of attack.

7.11 The expression "curvilinear" (or "nonlinear") regression can be used
in two different senses. In elementary texts, and even some advanced ones,
it refers to curvilinearity in the regressors x. But there are also situations
which are nonlinear in the coefficients β. For example, we might wish to
consider a model

$$y = \beta_0 + \gamma_1 e^{-\beta_1 x_1} + \gamma_2 e^{-\beta_2 x_2} + \epsilon$$

on the grounds that β_1 and β_2 are positive and y is known to tend to a constant for large x_1 and x_2.

Such models can, of course, also be fitted by least squares, but the resulting equations, if we differentiate with respect to the parameters in the usual way, are no longer of a simple form linear in those parameters. In general, a machine routine is required to proceed to a solution in an iterative way. Some methods for procedure are given in Beale (1969).

7.12   Recourse to the computer is also required when constraints are imposed on a regression situation, even if the standard linear form of (7.2) is concerned. Suppose, for example, that the external circumstances require that $\beta_1 \geqslant \beta_2 \geqslant \beta_3$, etc. We then have a problem in quadratic programming, namely to find the minimum of the quadratic form in β:

$$\Sigma \left(y - \sum_j \beta_j x_j\right)^2$$

subject to the linear constraints $\beta_1 - \beta_2 \geqslant 0, \beta_2 - \beta_3 \geqslant 0$, etc. In practice, perhaps, one would be inclined to estimate the unconstrained β's and hope that the inequalities were obeyed; but if this hope is not fulfilled we require the more sophisticated mathematical program.

7.13   There is one feature of data which occur in a temporal order that is worth examination. Suppose we have y and x, observed for a set of consecutive n years, and draw a scatter diagram. Let us then number the points in date order and join them up, one year to the succeeding year. It is sometimes found that a pattern emerges. Fig. 7.1 illustrates the point. The J-shaped curve shows, for the sequence of years 1861–1913, the rate of change of money wage rates (y) against unemployment x. The curve shown seems to represent reasonably well the average pattern of relationship, and makes economic sense in that higher unemployment meant less bargaining power on the part of labour.

However, if we number the points by year and join them up we find a rather different picture. Fig. 7.1 also shows one subset for the 12 years 1868–79; other subsets show the same kind of effect to a greater or less extent. The system seems to have described a series of oscillations, providing a loop of the kind that in engineering might be called a "hysteresis loop". Clearly the above-mentioned curvilinear relation is only an average of two different relations which were better kept distinct.

7.14   Situations of the foregoing kind are easy enough to deal with when only one regressor is involved, but for $p > 1$ it is hard to keep one's head

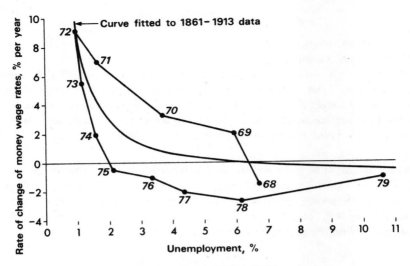

Fig. 7.1 (from Phillips, 1958)
By courtesy of the publishers of *Economica*

above the multiplicity of possibilities. On the whole it seems better to consider y against each x one by one to look for hysteresis effects, and, if they are found, to split each corresponding x into two variables. This can be done in at least two ways: (1) by calling x_1 the values in one half of the loop and x_2 the values in the other (x_1 being zero when x_2 is not and conversely); (2) by putting in another variable to express the rate of change of x with respect to time. The latter may involve the so-called autoregression system which is mentioned briefly at the end of this chapter, although again it is more properly dealt with as part of the analysis of time-series.

7.15   We turn now to a rather different class of problem, namely when some of the regressions are classificatory rather than continuous variables, and in the extreme case may be dichotomies. A particular case will make the general point. Consider y as the amount of a commodity bought by a consumer and x_1 as the price paid, and suppose that the purchasers may be male or female. It is tempting to represent sex as a $(0, 1)$ variable x_2 $(0 = \text{male}, 1 = \text{female})$ and to represent the purchase—price relationship by a single equation

$$y = \beta_0 + \beta_1 x_1 + \beta_2 x_2 + \epsilon. \tag{7.38}$$

Suppose there are n_1 males and n_2 females in the sample of n $(= n_1 + n_2)$ and let us denote the two classes by subscripts $_1$ and $_2$. Then the least-squares equations for estimating the β's are

$$\Sigma y - b_0 n - b_1 \Sigma x_1 - b_2 n_2 = 0$$
$$\Sigma y x_1 - b_0 \Sigma x_1 - b_1 \Sigma x_1^2 - b_2 \Sigma_2 x_1 = 0$$
$$\Sigma_2 y - b_0 n_2 - b_1 \Sigma_2 x_1 - b_2 n_2 = 0, \tag{7.39}$$

leading to

$$b_0 = \bar{y} - b_1 \bar{x}_1 - b_2 \bar{x}_2 \tag{7.40}$$

$$b_1 = \frac{n_1 \operatorname{cov}_1 (y, x_1) + n_2 \operatorname{cov}_2 (y, x_1)}{n_1 \operatorname{var}_1 x_1 + n_2 \operatorname{var}_2 x_1} \tag{7.41}$$

$$b_2 = \bar{y}_2 - b_1 (\bar{x}_1)_2 - b_0. \tag{7.42}$$

Perhaps we should have expected this result. The slope of the regression line is the weighted average of the lines we should have got for men and women separately. The constant term in (7.38) will be $\bar{y}_1 - b_1(\bar{x}_1)_1$ for $x_2 = 0$ and $\bar{y}_2 - b_1(\bar{x}_1)_2$ for $x_2 = 1$.

Thus our attempt to put the two relations into one equation (7.38) has merely resulted in an average. If there really is a sex-difference it is obviously desirable to keep the two regressions distinct and not to use the pseudo-variable (0, 1).

7.16 For *ordered* classifications into more than two groups a new complication arises. Consider a three-way classification, say, young (n_1), middle-aged (n_2), and old (n_3). Let us impose pseudo-variables u_1, u_2, u_3 which for the moment we will not specify. We measure them about their overall mean so that

$$n_1 u_1 + n_2 u_2 + n_3 u_3 = 0. \tag{7.43}$$

In equation (7.38) we will also measure y and x_1 from their overall means in the set of $n = n_1 + n_2 + n_3$ individuals. Thus the estimate of β_0 is zero and the other equations are

$$\Sigma y x_1 = b_1 \Sigma x_1^2 + b_2 \Sigma x_1 x_2$$
$$\Sigma y x_2 = b_1 \Sigma x_1 x_2 + b_2 \Sigma x_2^2, \tag{7.44}$$

giving

$$b_1 = \frac{\Sigma y x_1 \Sigma x_2^2 - \Sigma y x_2 \Sigma x_1 x_2}{\Sigma x_1^2 \Sigma x_2^2 - (\Sigma x_1 x_2)^2}. \tag{7.45}$$

Σ here refers to the whole set of n, and we have

$$\Sigma x_2^2 = n_1 u_1^2 + n_2 u_2^2 + n_3 u_3^2 \tag{7.46}$$

$$\Sigma y x_2 = u_1 \Sigma_1 y + u_2 \Sigma_2 y + u_3 \Sigma_3 y \tag{7.47}$$

$$\Sigma x_1 x_2 = u_1 \Sigma_1 x_1 + u_2 \Sigma_2 x_1 + u_3 \Sigma_3 x_1. \tag{7.48}$$

It is then plain that b_1 depends on the relative values of the u's. If, for example, we adopt a common convention and put u_1, u_2, u_3 equal to $-1, 0, 1$, we are imposing on the data the condition that the middle category, in some measurable sense, lies half-way between the other two.

7.17 Suppose now that we have an unordered categorization, e.g. that there are three classes by religious denomination, namely Christians, Jews and Moslems. The procedure which suggests itself here, if we must have just one regression equation, is to represent the first by a $(1, 0, 0)$ variable, the second by a $(0, 1, 0)$ variable and the third by a $(0, 0, 1)$ variable. If the corresponding variables are denoted by x_2, x_3, x_4, we have to consider

$$y = \beta_0 + \beta_1 x_1 + \beta_2 x_2 + \beta_3 x_3 + \beta_4 x_4. \tag{7.49}$$

In an obvious notation we then have

$$\Sigma y = b_0 n \quad + b_1 \Sigma x_1 \quad + b_2 n_1 \quad + b_3 n_2 \quad + b_4 n_3$$
$$\Sigma y x_1 = b_0 \Sigma x_1 + b_1 \Sigma x_1^2 \quad + b_2 \Sigma_1 x_1 + b_3 \Sigma_2 x_1 + b_4 \Sigma_3 x_1$$
$$\Sigma_1 y = n_1 b_0 \quad + b_1 \Sigma_1 x_1 + b_2 n_1$$
$$\Sigma_2 y = n_2 b_0 \quad + b_1 \Sigma_2 x_1 \quad + b_3 n_2$$
$$\Sigma_3 y = n_3 b_0 \quad + b_1 \Sigma_3 x_1 \quad\quad\quad + b_4 n_3. \tag{7.50}$$

These five equations are not independent, the first being the sum of the last three. We could, in fact, have made do with only two pseudo-variables, say $(1, 0)$ for Christians and $(0, 1)$ for Moslems, a value of 0 for both being equivalent to defining the individual as a Jew. But the argument is not affected.

We find

$$b_1 = \frac{\sum_3 \text{cov}\,(y x_1)}{\sum_3 \text{var}\,x_1}. \tag{7.51}$$

The coefficient b_0 is indeterminate, but we have regarded it as absorbed in the terms b_3, b_4, b_5. Once again we find that the pseudo-variable approach averages the slope of the regression lines.

7.18 There are occasions when the regressand y is also represented as a pseudo-variable. We shall revert to this case in Chapter 10 when discussing discrimination.

7.19 It should be plain that the use of pseudo-variables requires some care. They result in the averaging of relations which are often better kept distinct. However, there are times when we may admit them, especially when the data embody a number of polytomies and the numbers in the sub-cells determined by them become so small that no relation based on a sub-class would be reliable. We must also recognize that in one sense the continuous variable is only an extreme case of categorization, and regression relations are also a kind of average. However, the continuous variable *ipso facto* determines in a metrical sense the intervals between the infinitely fine categorization which it defines. If our ordered categories can be quantifiably ascertained (Example 9.5, the case of the Egyptian skulls, is a case in point) we can proceed with regressions in the ordinary way.

7.20 We now proceed to discuss one of the most serious problems which beset multivariate analysis, and regression analysis in particular, that of collinearity or approximate collinearity. We have already noticed that the inverse of the covariance matrix occurs naturally in a number of multivariate contexts and that its vanishing is a source of ambiguity and indeterminacy.

In the extreme case, when one variable is a linear function of the other, the coefficients in a regression equation which includes them all are indeterminate, for the covariance matrix of regressors then vanishes. If some of the eigenvalues of that matrix are small, its determinant (being the product of the eigenvalues) is also small, and the coefficients will be ill-determined. In fact, the coefficients in a regression, contrary to general belief, are not of any particular significance if such near-collinearities are present.

Example 7.1

In some studies of demand analysis, Stone (1945) considered the consumption of beer in the United Kingdom for the years 1920–1939 inclusive. He was interested in an equation of the type

$$\log q = a + b \log Q + c \log p + d \log \pi + f \log g + rt \log e, \quad (7.52)$$

where q = consumption (bulk barrels), Q = real income, p = retail price, π = cost-of-living index, g = specific gravity of beer, t = time, and e is the base of natural logarithms, $2 \cdot 718 \ldots$. Calling the logarithms (to base 10) of $Q, p, \ldots, t$, respectively, $x_1, x_2, \ldots, x_5$, we have the following correlation matrix:

$$
\begin{bmatrix}
1 \cdot 0 & -0 \cdot 610\,375 & -0 \cdot 660\,691 & -0 \cdot 507\,697 & 0 \cdot 918\,651 \\
 & 1 \cdot 0 & 0 \cdot 447\,714 & -0 \cdot 256\,291 & -0 \cdot 462\,810 \\
 & & 1 \cdot 0 & 0 \cdot 397\,888 & -0 \cdot 831\,054 \\
 & & & 1 \cdot 0 & -0 \cdot 649\,439 \\
 & & & & 1 \cdot 0
\end{bmatrix}
\quad (7.53)
$$

The variables are obviously highly correlated. A principal component analysis gives the following results:

Eigenvalues	Coefficients of				
	x_1	x_2	x_3	x_4	x_5
3·246 993	0·521 543	−0·312 055	−0·475 347	−0·324 548	0·547 109
1·275 281	−0·071 142	0·709 006	0·038 129	−0·694 304	0·093 475
0·385 919	−0·472 955	−0·216 049	0·824 664	−0·222 385	0·010 494
0·070 021	0·521 933	0·594 254	0·010 309	0·580 105	0·194 475
0·021 786	−0·476 266	−0·010 942	0·303 997	0·163 198	0·808 705

$$(7.54)$$

As usual, we retain more decimal places than the accuracy of the primary data would justify, in order to provide checks on the arithmetic.

The correlations of y (standardized to unit variance) with the x's are

$$-0.457\,536, \quad 0.031\,719, \quad 0.899\,223, \quad 0.601\,129, \quad -0.710\,155 \quad (7.55)$$

Two eigenvalues are very small, and we might neglect them without seriously impairing the situation. However, to neglect components is not the same thing as neglecting variables. If we decided to neglect the last component, which is heavily weighted on x_5, we might with reason neglect x_5 (which would mean that the time-effect is taken care of by the temporal elements in the other variables). But to neglect the fourth component imposes a problem, for it is about equally weighted on x_1, x_2 and x_4. Let us compare some possibilities. The regressions of $\log q$ (about its mean) are as follows:

All five variables:
$$y = 0.2819x_1 - 0.3094x_2 + 0.9784x_3 + 0.1233x_4 - 0.2309x_5$$
$$R^2 = 0.9896 \quad (7.56)$$

x_3, x_4, x_5 retained:
$$y = 1.2506x_3 + 0.5487x_4 + 0.6855x_5 \qquad R^2 = 0.9676 \quad (7.57)$$

x_1, x_2, x_3 retained:
$$y = -0.021\,25x_1 - 0.4722x_2 + 1.0966x_3 \qquad R^2 = 0.9808 \quad (7.58)$$

There is little to choose between the values of R^2. It is true that if we consider the complementary quantities $1 - R^2$, equivalent to proportionate residual variance, namely, 0.0104, 0.0324, 0.0192, they appear different. But they are all small.

It will be plain that in such circumstances the actual coefficients in the equation are virtually meaningless. Those of x_1, for example, may be 0.2819, zero, $-0.021\,25$.

Owing to the high connectivity among the variables, some can assume responsibility, so to speak, for others. *It is the equation as a whole, not the coefficients of individual terms, which is important.*

7.21   It is worth while stressing the point we have just made. A traditional use of regression equations, especially in econometrics, was to argue that, for example, the coefficient of x_1 might be regarded as an elasticity of demand with respect to real income. If all the other variables remain constant, a unit increase in x_1 will result in an increase of b in $\log q$. The point is that, owing to intercorrelation, an increase in x_1 is accompanied by changes in the other variables, and we cannot realistically regard them as held constant, except perhaps in the very short term.

7.22   It is customary in introductory texts dealing with regression to observe that the significance of a regression coefficient can be tested by the use of

Student's t. For example, with y regressed on a single variable x with coefficient b,

$$t = (b - \beta)\left\{\frac{(n-2)\,\mathrm{var}\,x}{s^2}\right\}^{\frac{1}{2}}, \tag{7.59}$$

where s^2 is the residual sum of squares, is distributed as t with $n-2$ degrees of freedom.

The same kind of test is often employed where y is regressed on $x_1 \ldots x_p$, i.e.

$$t = \frac{b_j - \beta_j}{\{s^2\,[xx']_{ii}^{-1}\}^{\frac{1}{2}}} \tag{7.60}$$

is tested with $n-p$ degrees of freedom. The logic of this procedure is not free from objection. In the ordinary linear model with p variables, the estimate of residual variance $s^2/(n-p)$ is distributed independent of any b, which is itself a linear function of y and therefore is normally distributed. If, however, we use the t-test to reject a coefficient, we are doing so on the basis of a residual variance calculated by not rejecting it. The criterion is obviously even more questionable if we reject several variables at one go.

7.23   The difficulties involved when the x's are correlated lead naturally to the question whether we can transform them to variables which are not correlated. In particular the principal components suggest themselves for this purpose. Something may indeed be gained, but not much.

Example 7.2

Consider again the data of Example 7.1. If we write ξ for the principal components, the regression is

$$y = \sum_{j=0}^{p} \alpha_j \xi_j + \epsilon. \tag{7.61}$$

In virtue of the orthogonality of the components, we have

$$\alpha_j = \frac{\sum y\xi_j}{\sum \xi_j^2} = \frac{\sum y\xi_j}{\lambda_j}. \tag{7.62}$$

Furthermore, the contribution to the variance of y due to the fitting of ξ_j is $\alpha_j^2 \lambda_j$.

For instance, in our data, taking the values of $\sum yx_j$ from (7.55), we have

$$\begin{aligned}
\alpha_1 &= (0{\cdot}521\,543)(-0{\cdot}457\,536) + (-0{\cdot}312\,055)(0{\cdot}031\,719) \\
&\quad + (-0{\cdot}475\,347)(0{\cdot}899\,223) + (-0{\cdot}324\,548)(0{\cdot}601\,129) \\
&\quad + (0{\cdot}547\,109)(-0{\cdot}710\,155) \\
&= -0{\cdot}387\,926.
\end{aligned}$$

Likewise,

$$\alpha_2 = -0.309\,293 \quad \alpha_3 = 0.977\,343 \quad \alpha_4 = -0.001\,048$$
$$\alpha_5 = 0.675\,664.$$

The contributions to the variance of y (which has been scaled to unity) are $\lambda_i \alpha_i^2$, $i = 1, \ldots, 5$, namely 0·4886, 0·1220, 0·3686, + 0·0000, 0·0099. The first three components account for 97·92 percent of the variance. However, since the ξ's depend on all the x's, even if we omit ξ_4 and ξ_5, we are still left with expressions depending on all the x's. We note also that the order of importance of contribution to the variances is not necessarily the same as the order of magnitude of the eigenvalues.

7.24 It is often more relevant, if economy is our aim, to discuss the rejection of variables. One possible way of proceeding is known as the *stepwise backwards* method. We work out the regression on all variables, giving us a multiple correlation of, say, R_p^2. We then examine which variable, on discard, lowers R_p^2 the least, to, say, R_{p-1}^2. We then examine which of the remaining variables can be discarded to give R_{p-2}^2 and so on. If the successive discarding process lowers R^2 very little until we reach R_{p-k}^2 then we have, on the face of it, safely discarded k variables, with very little sacrifice of information.

An alternative is to proceed *stepwise forward*. We regress y on each variable in turn and retain that one with the highest R_1^2. Then we ascertain which of the remaining $p - 1$ variables adds most to the multiple correlation and bring that in to give R_2^2; and so on until we reach a satisfactory value of R^2, whereupon we discard the remaining variables.

Unfortunately these methods suffer from two drawbacks. First, they may not give the same answer. Second, if they do, that answer may not be the optimum.

Example 7.3 (Oosterhoff, 1963)

In the paper by Oosterhoff under reference he gives a set of artificial observations which yield the following sums of squares in regression as a fraction of the total sums of squares:

x_1	0·6397	x_2, x_4	0·8138	x_1, x_2, x_3	0·9644
x_4	0·5608	x_1, x_2	0·7627	x_2, x_3, x_4	0·9144
x_2	0·2528	x_1, x_3	0·6899	x_1, x_2, x_4	0·8179
x_3	0·0906	x_1, x_4	0·6439	x_1, x_3, x_4	0·6906
		x_3, x_4	0·5608		
		x_2, x_3	0·2563		

Here, for example, the regression of y on x_1 yields a multiple correlation R^2 of 0·6397. That on x_1 and x_2 yields 0·7627, and so on. That on all four gives $R^2 = 0.9737$.

Proceeding stepwise backward, we should retain x_1, x_2, x_3, discarding x_4; then, of the pairs involving those three retain x_1 and x_2; and finally retain x_1, discarding x_4, x_3, x_2 in that order. A stepwise forward method would lead to the same result. Should we, then, have decided to retain two variables, that pair would have been x_1 and x_2. But it is seen that x_2, x_4 are a better pair.

Oosterhoff gives some even more telling examples. In one set involving 10 regressions the stepwise forward method involved the variables in the order $x_1, x_2, \ldots, x_{10}$. The sum of squares due to x_1 and x_2 was 0·2317. That for the first eight was 0·8852. But that for x_9 and x_{10} alone was 0·9830.

The following (imaginary data) give a set of 10 values of a regressand and two regressors:

y	x_1	x_2
29	7	7
−48	−19	−12
18	38	39
−12	45	49
44	−5	−7
57	15	12
47	−38	−40
10	38	39
86	59	53
46	−27	−29

If we calculate the correlation between y and x_1 we find a value of 0·104, and R^2 (which in this case is r^2) is 0·011. Likewise the correlation of y and x_2 is −0·006 35 and $R^2 = 0·000$. We might well conclude that there was no point in regressing y on x_1 and x_2 for it is almost independent of both.

However, x_1 and x_2 are highly correlated, with a correlation coefficient of 0·9938, and if we determine the regression of y on x_1 and x_2 we find that $R^2 = 0·999$. The regression is

$$y - \bar{y} = 8·948(x_1 - \bar{x}_1) - 8·898(x_2 - \bar{x}_2), \qquad (7.63)$$

where

$$\bar{y} = 27·7, \quad \bar{x}_1 = 11·3, \quad \bar{x}_2 = 11·1.$$

It is instructive to consider from the geometrical viewpoint how such a thing may happen. Let us represent the variables by vectors in our second type of space, as in Fig. 7.2. We imagine x_1 and x_2 as lying in the plane of the paper and y as nearly doing so. The angle between x_1 and x_2 is small, indicating high correlation between them. The angles between y and x_1, and between y and x_2 are nearly right-angles, indicating only slight correlation. But y lies almost in the (x_1, x_2) plane, which implies that it is nearly a linear function of x_1 and x_2. We see, in fact, from (7.63) that y is very highly dependent on the *difference* between x_1 and x_2.

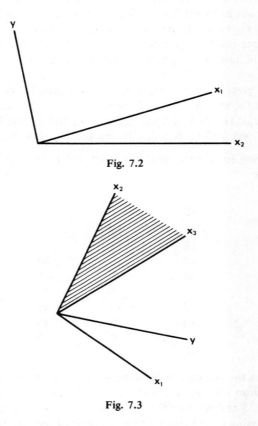

Fig. 7.2

Fig. 7.3

The same kind of representation enables us to understand how a "best" variable brought in by stepwise methods can be rejected as more variables are brought in. We can only draw a simple case, as in Fig. 7.3 which shows two variables x_2 and x_3 in the plane of the paper and x_1 outside that plane. The regressor y is also to be regarded as outside the plane of the paper but close to it. The angle between y and x_1 is imagined to be smaller than those between y and x_2 or between y and x_3. The best single variable is then x_1. But when we proceed to pairs of variables, y is closer to the (x_2, x_3) plane than to the (x_1, x_2) or (x_1, x_3) planes. Consequently the best pair of variables is (x_2, x_3). It is, however, conceivable (though difficult to draw) that if we proceeded to a fourth dimension with another variable x_4, y would be closer to the hyperplane (x_1, x_3, x_4) than to any of the others, and hence the best set would be (x_1, x_3, x_4); and so on.

7.25 Results such as this make us suspicious of stepwise methods. Given enough computational assistance, we should do better to work out all the possible regressions with their associated R^2 and choose from the complete array of possibilities. Any variable can be either in or out of the regression, so the total number of possibilities is 2^p (or rather $2^p - 1$, because we should not bother with the trivial case where no variable appears). Programs have been written for electronic computers which will work out all the regressions, but they are expensive in machine time for $p \geqslant 14$ (Garside, 1965). An alternative program (Beale *et al.*, 1967) achieves the optimum set and sets in the neighbourhood much more efficiently by developing cut-off rules to avoid the calculation of sub-optimum sets. Other procedures are described in a review article by Draper and Smith (1970).

7.26 A good program will then list, with associated R^2, the best single variable, the best pair, the best triad, . . . , the best (and only) set of p, and sets which are "nearly" the best in some tolerance sense. Which set is chosen depends on how satisfactory we regard the associated R^2.

It does not follow that the best set of k includes all the variables in the best set of $k - 1$. Sometimes a variable is discarded at one stage and brought back later for a set of different size. Where there is general instability of this kind, the optimal sets fluctuating as we go from k to $k + 1$ to $k + 2$, etc., the situation needs more penetrating analysis in regard to the eigenvalues and the original data. But cases of this kind are rare.

Example 7.4

Criteria of discard by reference to correlations themselves are dangerous. Consider the case of a medical man concerned with conditions of the spine such as displaced vertebrae. Measurements which he might take on the body include the length of the legs. Now the length of one leg is so highly correlated with the length of the other that it might be regarded as a waste of time to measure both. But if we reject one variable as unnecessary we should miss an important contributor to spinal deformation, the difference of leg lengths.

7.27 We can now pick up a point which had to be left aside in Chapter 2, namely the possibility of discarding variables in principal component analysis. One obvious procedure is to find the eigenvectors of the covariance matrix, and if one corresponding eigenvalue is near to zero, then equate that eigenvector to zero and express one of the variables in terms of the others. Strictly speaking, we should then calculate the matrix of $p - 1$ variables and repeat the procedure; and so on until we winnow the variables down to a group with no small eigenvalues. This at least gets rid of collinearities, but it takes a good deal of time and is subject to one other disadvantage, namely that if

$\Sigma\,l_j x_j$ is small, we do not know which variable to substitute. (The argument that it makes no difference is not convincing in arithmetical practice.)

7.28 A preferable procedure was developed by Beale *et al.* (1967) from a modification of the program dealing with optimal regressions. The basis of the procedure is the consideration that we reject a variable if it has a high R^2 when regressed on the other variables; the point being that if this is so we can very approximately regenerate the discarded variables from the retained ones. Likewise, we can reject a set of k variables if the least R^2 on the retained set exceeds some acceptable defined value; for the worst we can do is to lose the proportion $1 - R^2$ of the information on the k rejects.

Example 7.5

Seventeen measurements are given on 67 lubricating oil basestocks. This material was thought to be homogeneous and not to contain any odd or outlying observations. There are no missing values.

The 17 variables are as follows:

(1) Kinematic viscosity at $100°F$.
(2) Kinematic viscosity at $140°F$.
(3) Kinematic viscosity at $210°F$.
(4) Viscosity index.
(5) Refractive index.
(6) Molecular weight.
(7) Sulphur (percent).
(8) Specific gravity $60°F/80°F$.
(9) Flash-point
(10) Pour-point.
(11) Neutralization value mg KOH/g.
(12) Carbon residue.
(13) Percent N + P (weight percent naphthalenes plus paraffin).
(14) Percent A + S (weight percent aromatic plus sulphur compounds and other polar materials).
(15) Percent C_A (percent carbon in aromatic rings).
(16) Percent C_N (percent in naphthalene rings).
(17) Percent C_p (percent not in ring structure, i.e. in paraffin + alkyl groups).

These data are not linearly independent. In fact, variables 13 and 14 always add up to 100, and variables 15, 16 and 17 always add up to 100. Variables 1, 2 and 3 are also nearly linearly dependent, log viscosity being a linear function of temperature. No special provision was made in the analysis for these dependences.

The first column and the main body of Table 7.1 are self-explanatory. In the penultimate column is shown the least value obtained by regressing each

Table 7.1 *Interdependence analysis of lubricating oil basestock measurements*

No. of variables selected	Variables selected															Min R^2 with rejected variables	Max R^2 between selected variables
1	16	–	–	–	–	–	–	–	–	–	–	–	–	–	–	0·0177	–
2	6	12	–	–	–	–	–	–	–	–	–	–	–	–	–	0·2010	0·4367
3	4	5	11	–	–	–	–	–	–	–	–	–	–	–	–	0·4722	0·6999
4	1	2	4	5	–	–	–	–	–	–	–	–	–	–	–	0·5092	0·9800
5	3	7	10	11	17	–	–	–	–	–	–	–	–	–	–	0·7327	0·4256
6	3	5	7	10	11	17	–	–	–	–	–	–	–	–	–	0·8077	0·9094
7	7	8	9	10	11	12	16	–	–	–	–	–	–	–	–	0·8591	0·6674
8	5	7	9	10	11	12	13	16	–	–	–	–	–	–	–	0·8895	0·8548
9	3	5	7	9	10	11	12	13	16	–	–	–	–	–	–	0·9527	0·9014
10	3	7	8	9	10	11	12	13	15	16	–	–	–	–	–	0·9605	0·9322
11	1	6	7	8	9	10	11	12	13	16	17	–	–	–	–	0·9655	0·9640
12	2	4	6	7	8	9	10	11	12	13	15	16	–	–	–	0·9869	0·9644
13	2	4	5	6	7	8	9	10	11	12	13	15	16	–	–	0·9931	0·9869
14	1	3	4	5	6	7	8	9	10	11	12	13	15	16	–	0·9997	0·9874
15	1	2	3	4	5	6	7	8	9	10	11	12	13	15	16	1·0000	0·9997

rejected variable on the retained variables. Thus, for example, in row 9 each of the six rejected variables has at least a value of R^2 equal to 0·9527 when regressed on the other nine. We see that nothing very useful can be done with less than five variables, but that six variables can account for over 80 percent of the variability in each of the nine rejected variables, nine can account for over 95 percent, and 13 for over 99 percent. We might well be content to cut the variables to nine, the variables selected for retention being listed in the table.

The last column may be of some interest: it gives the largest value of R^2 obtained by regressing one of the *selected* variables on the others, and indicates that the selected variables are usually not highly correlated among themselves. With less than nine variables selected, only on two occasions did this number exceed 90 percent.

It is of some interest to compare these results with those obtained by the use of eigenvectors. A component analysis carried out on the correlation matrix for all variables revealed two zero eigenvalues associated with the linear relationships noted earlier, and one eigenvalue of 0·000 16 associated with the nearly linear relationship between variables 1, 2 and 3. At this point variables 2, 14 and 17 were eliminated from the problem. A further component analysis was carried out on the correlation matrix for the remaining 14 variables. This produced four small eigenvalues of 0·000 45, 0·000 73, 0·001 60 and 0·002 06. It was then decided to eliminate four more variables associated with the eigenvectors corresponding to these eigenvalues. The principle adopted was to eliminate the variable with the largest coefficient in the vector expressed as a linear function of the variables. The variables removed in this way were 3, 4, 5 and 6, which left 1, 7, 8, 9, 10, 11, 12, 13, 15 and 16. This selection gives a minimum R^2 with the rejected variable of 0·9501, which is nearly as good as the optimum set. Judged by the criterion of the minimum R^2 with an unselected variable, 10 is not a particularly useful number of variables to select, and indeed the optimum program does marginally better with nine variables than the component analysis program with 10. A third component analysis was carried out on the correlation matrix for the 10 remaining variables. This produced three small eigenvalues of 0·003 72, 0·006 53 and 0·009 49 which were associated with the variables 1, 8 and 13, leaving 7, 9, 10, 11, 12 and 15. This selection gives a minimum R^2 with the rejected variables of 0·8531 and an average R^2 of 0·9078, as compared with 0·8591 and 0·9068 for our program. There is therefore not much to choose between these selections.

7.29   Cases sometimes occur in which a regression changes at some point. If that point is identifiable in terms of the x's no great problem arises. More difficult is the case when the point of change itself has to be estimated, especially when more than one regression is involved. Most of the work which has been done in this field concerns curves in two dimensions. Even in this relatively simple case the situation has been imperfectly explored.

In some earlier work Quandt (1960) considered tests of the hypothesis that a linear regression follows two regimes. Robinson (1964) and Hudson (1966) considered the fitting of segmented curves when the join points have to be estimated. Least-squares criteria were used, but they are not simple to apply: the overall model is not linear — for example, if two lines

meet at $x = d$,
$$y = \beta_{10} + \beta_{11}x, \quad y = \beta_{20} + \beta_{21}x$$
$$\beta_{10} + \beta_{11}d = \beta_{20} + \beta_{21}d. \tag{7.64}$$

Bellman and Roth (1969) represent a curve by fitting a set of straight lines, proceeding by dynamic programming, and minimizing the absolute deviation. McGee and Carleton (1970) cluster the points and fit regressions piecewise.

7.30   It seems that in full generality the problem merits further attention. If there are an unknown number of points of change, unknown values where change takes place, and perhaps an unknown number of regressions, it obviously requires some assumptions to make the problem determinate — even in two dimensions we can join a set of n points one to the next and obtain $n - 1$ "perfect" regressions. The best procedure in practice, perhaps, is to guess at the number of points of change, from inspection of scatter diagrams or otherwise, and to decide, again by inspection, for the majority of points, which part of the variation they fall into. There remain a number of points near the joins which might fall on either side of the join, so to speak. From then on they can be allocated by exhausting all possibilities and choosing the one that minimizes the sum of squares of deviations over all segments of the regression. If (as in the next chapter) the x's are subject to observational error, a new type of difficulty appears, and in the present state of knowledge practical solutions can only be found in a very pragmatic way.

7.31   Not infrequently we have a set of n observations on $y, x_1, \ldots, x_p$, in which some values are missing. It is desirable to have a systematic method or set of methods for dealing with such a situation.

If the number of imperfect sets is small relative to n, it may be better to reject them and work solely with the sets for which all observations are available. One must, of course, have regard to the reasons why the observations are missing and be reasonably sure that the complete set are a random sample or are relevant to our inquiry. (A study of the curative effects of certain treatments would hardly be satisfactory if we omitted patients who had died under it simply because we could not continue to take measurements on them.)

7.32   If, however, there are a large number of imperfect records, i.e. individuals for which some but not all the values of the x's are known, there is a strong inducement not to reject them entirely; to do so sacrifices a lot of information. We must always remember that there is no substitute for knowledge. To replace the missing values by estimated quantities does not add anything to the body of information. It does, however, simplify our analysis by enabling the completed data to be put through a standard routine.

7.33   Various methods have been proposed to replace missing values by estimated quantities. They all depend on using completed or partially completed records to estimate the unknown values by a regression technique or a covariance technique. For example, we can work out the covariance of each pair of x's, and the covariance of y with each x, from such data as are available and then substitute these estimates in the normal equations such as (7.3). If the number of missing values is relatively large, however, this may lead to serious biases. The matter was studied by Haitovsky (1968) in some detail by a simulation technique.

Example 7.6 (Haitovsky, 1968)

One hundred observations were generated according to

$$y = 150 \cdot 0 + 5 \cdot 0x_1 - 2 \cdot 0x_2 + 0 \cdot 3x_3 + 3 \cdot 0x_4. \qquad (7.65)$$

The observations were multivariate normal and were generated so as to have the correlation matrix

y	x_1	x_2	x_3	x_4	
1·0	0·7852	0·6137	0·6389	0·8333	
	1·0	0·8738	0·5166	0·4267	
		1·0	0·6314	0·4650	
			1·0	0·7119	
				1·0	(7.66)

The method of ordinary least squares (OLS) was applied to all 100 observations. Then some were rejected at random, i.e. 6 y's, 25 x_1, 15 x_2, 0 x_3 and 10 x_4. The regression was calculated for the remainder. This was repeated 10 times, with the same number of rejects but different rejected individuals. This is called Method 1.

The covariances were then estimated from the available data (after rejection) and the normal equations solved. This was repeated seven times. This is called Method 2, but there was selective choice of x_1, some being chosen from the larger values.

The results, with the average of the 10 or seven runs as the case may be, were as follows:

	Constant	x_1	x_2	x_3	x_4	
True	150·0	5·0	−2·0	0·3	3·0	
OLS	150·732	4·968	−1·922	0·514	2·922	($R^2 = 0.992$)
Method 1	147·327	5·006	−5·968	0·775	2·907	
Method 2	414·443	4·116	−0·660	−6·582	2·699	

The bias introduced by Method 2 is obvious, and in this example it is clearly better to ignore the imperfect sets, notwithstanding that they amount to 25 percent of the whole. This, in fact, was Haitovsky's general conclusion.

7.34   The subject has been reviewed by Beale and Little (1973) who discuss six different approaches. On the basis partly of theory and partly of simulation studies they conclude that the best procedure is a modification of one due to Buck (1960) which they describe as Modified Maximum Likelihood. We begin by using the complete observations to estimate the means of all the variables and the covariance matrix. These values can then be used to estimate any of the missing quantities as linear functions of the variables which are known for a particular observation. These estimates are then substituted for the unknown values and a revised estimate constructed of means and covariances. The process is repeated until there is convergence. The point of the technique is that it is not necessary to calculate the means and covariances *ab initio* from the primary complete plus estimated observations; the previously estimated means and covariances can be adjusted to take account of the new estimates. The determination of these adjustments is a rather subtle matter and by no means as straightforward as it looks. The procedure was considered by Buck (1960) and by Orchard and Woodbury (1972).

7.35 Finally, we may briefly refer to the type of relationship known as autoregression. If observations are taken at a series of equal time-intervals, we are led to consider the relation of y at time t to the values of y at time $t-1$, $t-2$, etc., as well as on x's which may also have time subscripts; for example

$$y_t = \alpha_0 + \alpha_1 y_{t-1} + \ldots + \alpha_k y_{t-k} + \beta_1 x_{1t} + \beta_2 x_{1, t-1} + \ldots$$
$$+ \gamma_1 x_{2t} + \gamma_2 x_{2, t-1} + \text{etc.} + \epsilon_t. \tag{7.67}$$

The theory of such expressions belongs to Time-series Analysis and is dealt with in Kendall (1973). We mention it here to warn the reader that the standard theory of regression is not applicable to such equations. The way in which y appears partly as regressor and partly as regressand impairs the customary least-squares properties of regression coefficients.

A theorem of Mann and Wald (1943) states, in effect, that classical theory applies for large samples, but little is known for small or medium sample sizes.

NOTES

(1)   Geary (1970) considers the testing of randomness in residuals by counting the number of changes of sign. The distributional properties of this number are difficult to derive, but from a comparison on artificial and real data Geary concluded that the count of sign changes, considered as a binomial with parameter $\frac{1}{2}$, is comparable in average efficiency with the Durbin–Watson test.

(2)   Vinod (1973) has extended the Durbin–Watson test to some extent. If the first serial correlation of residuals is accepted as zero, it is possible to test the second against zero conditionally on the first being zero; and so on.

(3)   Some attention has been given to what is known as "ridge regression" methods, as to which see Draper (1963).

Where a matrix of covariances is ill-conditioned it may be brought nearer to stability (i.e. its determinant is further from zero) by adding a positive constant to the diagonal terms. The behaviour of the matrix under variation in this constant is then used to decide on an appropriate choice of that constant. It is not plain to me that this admitted distortion of the data (the effect of which is to diminish the correlations among the variables) has any theoretical justification.

(4)   When the members to which the regression is fitted are ordered at equal intervals, as in time-series analysis, there may be a case for fitting the regression to first or second differences. But it is open to doubt whether this method is very efficient. See Geary (1972) and Kendall (1973).

(5)   In some subjects, for example econometrics, there is also doubt whether the residuals, even if independent, are normally distributed. The question arises whether it would be better to fit, not by least squares, but by minimizing the sum of other powers of the residuals, e.g. the absolute deviations. This is now relatively easy on the computer. The theory, however, becomes rather complicated. It may nevertheless be necessary to reconsider much of what has been done by least squares in the light of the alternative possibilities.

(6)   For some further discussion of the problem of dummy variables see Kendall, M. G. (1979).

(7)   Attempts to overcome the problem of apportioning the relative influence of regressors on regressand have led to what is known as path analysis. For an introductory account of this subject see Kendall, M. G. and O'Muircheataigh, C. (1977).

8

Functional relationship

8.1 Suppose a physicist sets out to investigate the relationship between pressure P and volume V in a gas. By a few primitive experiments he is led to suspect a relation of the Boyle type, $PV = $ constant, or equivalently,

$$\log P + \log V - \log K = 0, \tag{8.1}$$

where P and V are respectively pressure and volume and K is a constant. He may then wish to verify the law more closely by taking readings over a range of values of P and V and over time. If he plots the variables $\log P$ and $\log V$ on ordinary graph paper he will get a set of points lying nearly on a straight line, and the best line he can draw will give him an estimate of the constant K. The problem as to what is "best" arises because the points do not lie on a straight line exactly.

8.2 This simple situation enables us to make a point which is often ignored in the treatment of statistical data: the method of analysis depends on the model we have in mind; and this may depend on the way in which the data were obtained. In this particular case there are at least four ways of looking at the situation.

(1) We conduct an experiment in which we assign values of P *without error* and measure the values of V (with error). The model is then

$$\log V = \beta \log P + \log K_1 + \epsilon_1. \tag{8.2}$$

This is a regression situation, $\log V$ being the regressand and $\log P$ the regressor. We can, if desired, restrict the model by imposing the constraint that $\beta = -1$.

(2) We may experiment by adjusting V without error and observing P with error. The roles in (8.2) are then inverted to give

$$\log P = \gamma \log V + \log K_2 + \epsilon_2. \tag{8.3}$$

If we are given observations but do not know which of (8.2) or (8.3) applies, we might analyse by both models and hence get two different

109

answers. The statistician has been criticized for producing two regression equations in such circumstances, but the fault (if it is such) really lies in the imperfect specification of the model.

(3)   More realistically, perhaps, we may regard (8.1) as a true law of behaviour but contemplate errors in both P and V. The model is now that we observe p and v, say, given by

$$p = P + \epsilon \qquad (8.4)$$
$$v = V + \eta, \qquad (8.5)$$

where ϵ and η are random variables. This is *not* a regression situation, but one of functional relationship between variables subject to observational error. There are, of course, various minor modifications which we can make in such a model, for example, in supposing that ϵ and P are not independent (bigger errors in the larger quantities) or that errors in ϵ and η are correlated.

(4)   We may suspect that (8.1) is not complete and that we have left out part of the true model (as, in this case, we have, namely the temperature T and the effect of adiabatic expansion). We may then consider a model

$$\log P - \gamma \log V - \log K = \zeta, \qquad (8.6)$$

where ζ represents the omission and stands for something-left-out which we hope behaves more or less like a random error, so that regression methods can be applied. This is a fairly common model in the behavioural sciences.

8.3   For the most part in this chapter we are concerned with case (3), that is to say, the case in which the underlying relationship is functional but estimational problems arise because of errors of observation. In the first place we will consider the simplest case, when two variables Y and X are connected by the relation

$$Y = \alpha_0 + \alpha_1 X. \qquad (8.7)$$

The observed quantities ξ and η are given by

$$\left. \begin{array}{l} \xi_i = X_i + \delta_i \\[4pt] \eta_i = Y_i + \epsilon_i \end{array} \right\} \quad i = 1, 2, \ldots, n. \qquad \begin{array}{l} (8.8) \\[8pt] (8.9) \end{array}$$

We assume further that δ and ϵ are uncorrelated with each other and from one value of i to another; and that they have zero mean. We let

$$\operatorname{var} \delta_i = \sigma_\delta^2 \qquad (8.10)$$
$$\operatorname{var} \epsilon_i = \sigma_\epsilon^2. \qquad (8.11)$$

If we substitute from (8.8) and (8.9) in (8.7) we obtain

$$\eta = \alpha_0 + \alpha_1 \xi + (\epsilon - \alpha_1 \delta). \tag{8.12}$$

This illustrates our remark about the situation not being one of regression. The "residual" $\epsilon - \alpha_1 \delta$ is not independent of ξ, for both are random variables and

$$\text{cov} (\xi, \epsilon - \alpha_1 \delta) = \text{E} (X + \delta)(\epsilon - \alpha_1 \delta)$$
$$= -\alpha_1 \text{ var } \delta. \tag{8.13}$$

Equation (8.12) is a *structural* relation among random variables, as distinct from (8.7) which is a *functional* relation among mathematical variables.

8.4 Before passing on to consider (8.12) in detail we may notice one case in which the estimation of α_0 and α_1 can be dealt with by regression methods. This is Berkson's case (1950) of controlled variables.

Suppose we are conducting an experiment in which we control the variable X (or think we do) and observe Y with error, namely as η. In fact we specify ξ at various points, but in actuality our true value, purporting to be ξ, is X with error $\xi - X = \delta$. Thus the real control values are the unknown X's and

$$\xi = X + \delta,$$

where X is now the random variable, perfectly negatively correlated with δ because ξ is fixed. We then have $X = \xi - \delta$, and on substitution in (8.7),

$$\eta = \alpha_0 + \alpha_1 \xi + (\epsilon - \alpha_1 \delta), \tag{8.14}$$

which is of exactly the same form as (8.12) but fundamentally different because ξ is not a random variable and neither ϵ nor δ is correlated with it. Thus (8.14) is an ordinary regression and α_0, α_1 can be determined by least squares. We can also estimate the variance of $\epsilon - \alpha_1 \delta$ but not the variances of ϵ and δ separately.

8.5 The reader encountering controlled variables for the first time may find the foregoing argument a little subtle. It has the baffling property of certain kinds of optical illusions — suddenly jumping from one appearance to another. The essential point is that by deciding on the controlled values ξ, even if we do not exactly achieve them, we throw the random element from ξ to X. Provided that ϵ and δ are uncorrelated with ξ, ordinary least-squares methods apply. There are many experimental situations where this will be so. The dilemma of two regressions is resolved by destroying the symmetry between the variables. Only one of them can be the controlled variable.

8.6 A line of argument for the general case was advanced by Geary (1949). It has the advantage of not requiring any further assumption, but the severe practical disadvantage of not working for normally distributed errors and of giving highly unstable results for moderate non-normality. As, moreover, it requires some results in the theory of cumulants with which the reader may be unfamiliar we will treat it rather briefly.

A relation such as (8.7) or a more general linear relation can be put in the homogeneous form

$$\alpha_1 X_1 + \alpha_2 X_2 + \ldots + \alpha_k X_k = 0. \tag{8.15}$$

If each X_i has an associated error term δ_i and all δ's are independent of X's and of one another, the product-cumulants of the X's are the same as those of $\xi = X + \delta$, and for any set of integral p's,

$$\kappa_X(p_1, p_2, \ldots, p_k) = \kappa_\xi(p_1, p_2, \ldots, p_k), \tag{8.16}$$

provided that at least two of the p's are not zero. It may be shown that a consequence of (8.15) is a linear relation in the cumulants

$$\alpha_1 \kappa(p_1 + 1, p_2, \ldots, p_k) + \alpha_2 \kappa(p_1, p_2 + 1, \ldots, p_k) + \ldots$$
$$+ \alpha_k \kappa(p_1, p_2, \ldots, p_k + 1) = 0. \tag{8.17}$$

Hopefully, then, one could use equations of type (8.17) to estimate the α's. Unfortunately,

(1) If the variables ξ are normal all the product-cumulants vanish and the equations disappear.

(2) If the distributions are symmetrical but not normal the odd-order cumulants vanish.

(3) In other cases there is an unlimited suite of equations (but we choose those with low-order cumulants as being less subject to sampling error).

Example 8.1

Consider the case of (8.7). We will measure about the means of ξ and η, in which case α_0 is estimated as zero and the relationship may be written in the form

$$\alpha_1 X - Y = 0. \tag{8.18}$$

The equations (8.17) then become

$$\alpha_1 \kappa(p_1 + 1, p_2) - \kappa(p_1, p_2 + 1) = 0. \tag{8.19}$$

This is nugatory for normal variation and to avoid problems of symmetry we require $p_1 + p_2 + 1$ to be even. Thus the simplest cases are

$$\left.\begin{array}{ll} p_1 = 1, \ p_2 = 2, & \alpha_1 = \kappa_{13}/\kappa_{22}, \\ p_1 = 2, \ p_2 = 1, & \alpha_1 = \kappa_{22}/\kappa_{31}. \end{array}\right\} \tag{8.20}$$

8.7   The controlled variable case can be extended to curvilinear relationships.

Suppose we have the underlying relation

$$Y = \alpha_0 + \alpha_1 X + \alpha_2 X^2 + \alpha_3 X^3. \tag{8.21}$$

Writing $\xi - \delta$ for X, we have

$$\eta = \alpha_0 + \alpha_1(\xi - \delta) + \alpha_2(\xi - \delta)^2 + \alpha_3(\xi - \delta)^3 + \epsilon. \tag{8.22}$$

In this formulation δ is independent of the controlled variable ξ; so, on multiplying (8.22) by 1, ξ, ξ^2, ξ^3 in turn and taking expectations, we get

$$E(\eta) = \alpha_0 + \alpha_1 E(\xi) + \alpha_2\{E(\xi^2) + E(\delta^2)\} + \alpha_3 E(\xi^3), \tag{8.23}$$

and three analogous equations. It will simplify the exposition and lose no real generality if we suppose that the scale of ξ and the points of observation have been chosen so that odd order moments of ξ vanish. (This is particularly easy if ξ is observed at an odd number of equal intervals.) Writing μ_2, μ_4 for the moments of δ, we then get

$$E(\eta) \;\; = \alpha_0 \qquad\qquad + \alpha_2(\mu_2 + \operatorname{var}\delta) \tag{8.24}$$

$$E(\eta\xi) \;\; = \qquad \alpha_1\mu_2 \qquad\qquad\qquad + \alpha_3(\mu_4 + 3\mu_2 \operatorname{var}\delta) \tag{8.25}$$

$$E(\eta\xi^2) = \alpha_0\mu_2 \;\; + \alpha_2(\mu_4 + \mu_2 \operatorname{var}\delta) \tag{8.26}$$

$$E(\eta\xi^3) = \qquad \alpha_1\mu_4 \qquad\qquad\qquad + \alpha_3(\mu_6 - 3\mu_4 \operatorname{var}\delta). \tag{8.27}$$

The expectations on the left can be estimated from observed quantities. We have four equations in the constants α, but they incorporate var δ which appears as a nuisance parameter.

μ_2 times (8.24) minus (8.26) gives us an estimator of α_2. Likewise μ_4 times (8.25) minus μ_2 times (8.27) gives us an estimator of α_3. We therefore reach the peculiar situation in which we can estimate the curvilinear terms corresponding to α_2 and α_3, but not the terms α_0 and α_1 without a knowledge of var δ. There seems no way of estimating var δ except by replicating the experiment *with the same values of the controlled variables*.

It may be added that, in distinction to the linear situation of (8.12), where we can estimate the residual variance of $-\alpha_1\delta + \epsilon$, the error variance in (8.22) varies from one set of ξ's to another, and so no estimate of experimental error is obtainable and no ordinary tests of significance of the estimators of α are available.

8.8 Geary's method of **8.6** can also be extended to curvilinear relations, but under the same restrictions that the equations disappear for normal distribution and are unreliable near normality. In any case the equations, though linear in the parameters, are nonlinear in the cumulants. Reference may be made to Kendall and Stuart, vol. 2, chapter 29 for details.

8.9 The indeterminacy of the situation, even in the simple linear case, affects most of the methods to be discussed in this chapter. Before proceeding to a more orthodox approach, we consider two further methods of attacking the problem, the use of instrumental variables and of ranking.

8.10   Suppose that, when observing η and ξ, we also observe a third variable ζ which is correlated with the unobservable true value X but not with the errors of observation. ζ is then called an "instrumental variable" if it is used as an instrument in the estimation of the relationship between Y and X. Measuring the observables from their respective means, let us consider a_1 as an estimator of α_1 in (8.7):

$$a_1 = \sum_{i=1}^{n} \zeta_i \eta_i \bigg/ \sum_{i=1}^{n} \zeta_i \xi_i. \tag{8.28}$$

Write this as

$$a_1 \Sigma \zeta_i \xi_i - \Sigma \zeta_i \eta_i = 0$$

which, on substitution, becomes

$$a_1 \Sigma \{\zeta_i(X_i + \delta_i)\} - \Sigma \{\zeta_i(\alpha_0 + \alpha_1 X_i + \epsilon_i)\} = 0. \tag{8.29}$$

ζ is by hypothesis uncorrelated with δ and ϵ, and for large n we shall have

$$a_1 \operatorname{cov}(\zeta, X) = \alpha_1 \operatorname{cov}(\zeta, X). \tag{8.30}$$

So long as the covariance of ζ and X is not zero, a_1 then tends in large samples to α_1 and is a consistent estimator.

8.11   The practical difficulty of employing instrumental variables is that we rarely know enough about the system to be sure that a particular random variable, though correlated with X, is uncorrelated with the errors ϵ and η. There is, however, one rather artificial instrumental variable which can usually be supposed to satisfy the necessary conditions to some extent, namely a grouping variable.

Suppose that we divide the observations into two equal-sized groups. Let $\bar{\xi}$ be the mean of ξ in the first group and $\bar{\xi}'$ that in the second group, and assign an instrumental variable $\zeta = +1$ to the members of the first group and -1 to those of the second. The estimator a_1 of (8.28) then becomes simply

$$a_1 = \frac{\bar{\eta}' - \bar{\eta}}{\bar{\xi}' - \bar{\xi}}. \tag{8.31}$$

This method is equivalent to dividing the points (ξ, η) into two groups and estimating the linear relationship by joining the centres of gravity of the group.

8.12   The method is a plausible distribution-free procedure, but even this is not free from difficulty. For (8.31) to be valid it is necessary that in the limit $\bar{\xi} - \bar{\xi}'$ does not equal zero (as it would if we divided the sample into two groups at random). The obvious course in dividing into groups is to choose all those with smaller values of ξ for one group and the remaining group for the other. But in this case the estimator a_1 is not consistent. In fact, the group of smaller values of ξ will tend to include those members with smaller errors (for larger errors would take them into the other group), so that the condition of independence of error and variate value is violated. I am bound to admit that if the errors are relatively small and the variables fairly well spread out I regard this as a rather precious point.

8.13 In the model of (8.7) we may also estimate α_0 as

$$a_0 = (\bar{\eta}_1' + \bar{\eta}) - a_1(\bar{\xi}' + \bar{\xi}) \tag{8.32}$$

and the error variances can also be estimated. For

$$\text{cov}(\xi, \eta) = \text{E}(X + \delta)(\alpha_0 + \alpha_1 X + \epsilon) = \text{cov}(X, Y)$$

$$= \alpha_1 \text{var} X \tag{8.33}$$

and

$$\text{var}\,\xi = \text{var}\,X + \text{var}\,\delta, \tag{8.34}$$

so that

$$\text{var}\,\delta = \text{var}\,\xi - \frac{\text{cov}(\xi, \eta)}{\alpha_1} \tag{8.35}$$

and can be estimated accordingly. Likewise

$$\text{var}\,\epsilon = \text{var}\,\eta - \alpha_1 \text{cov}(\xi, \eta). \tag{8.36}$$

Example 8.2

R.L. Brown (1957) gave nine pairs of observations generated from a true linear functional relationship $Y - \bar{\eta} = 2(X - \xi)$ and random normal errors δ and ϵ with common variance unity. The values were

$$1{\cdot}8 \quad 4{\cdot}1 \quad 5{\cdot}8 \quad 7{\cdot}5 \quad 9{\cdot}3 \quad 10{\cdot}6 \quad 13{\cdot}4 \quad 14{\cdot}7 \quad 18{\cdot}9, \quad \bar{\xi} = 9{\cdot}57$$

$$6{\cdot}9 \quad 12{\cdot}5 \quad 20{\cdot}0 \quad 15{\cdot}7 \quad 24{\cdot}9 \quad 23{\cdot}4 \quad 30{\cdot}2 \quad 35{\cdot}6 \quad 39{\cdot}1, \quad \bar{\eta} = 23{\cdot}14.$$

We omit the middle value so as to obtain two groups of equal size. We find for the two half-samples that

$$\bar{\xi} = 4{\cdot}800, \quad \bar{\xi}' = 14{\cdot}406, \quad \bar{\eta} = 13{\cdot}775, \quad \bar{\eta}' = 32{\cdot}075,$$

so

$$a_1 = \frac{32{\cdot}075 - 13{\cdot}775}{14{\cdot}406 - 4{\cdot}800} = 1{\cdot}91,$$

which, on such small samples, is a fairly good estimate of the true value 2. For the estimated variances and covariances of the eight observations, using a divisor $n - 1$, we find

$$\text{est var}\,\xi = 29{\cdot}735, \quad \text{est var}\,\eta = 112{\cdot}709, \quad \text{est cov}(\xi, \eta) = 56{\cdot}764,$$

so for the estimates of error variance

$$\text{est var } \delta = -0.054$$

$$\text{est var } \epsilon = 5.16.$$

These are very bad estimates, the true value being unity. The estimate of var δ in fact is negative.

These inaccuracies in the estimates of error variance are understandable and need not undermine our confidence in the estimate of α_1. On the contrary, if the true values of X and Y are widely spaced relatively to the errors (as here), the observed ξ and η are highly correlated (here $r = 0.98$), the two regression lines are close together, and a_1 will be close to both the regression of ξ on η, cov $(\xi, \eta)/\text{var } \xi$, and the reciprocal of the regression of η on ξ, var $\eta/\text{cov }(\xi, \eta)$. Thus from (8.35) and (8.36) the estimates of var δ and var ϵ are highly unstable, and quite small deviations of a_1 from the true value α_1 may distort them violently.

8.14   It was pointed out by Nair and Shrivastava (1942) and Bartlett (1949) that the efficiency of the grouping method may be increased by dividing the sample into three groups and estimating α_1 by joining the centres of gravity of the two extreme groups. The same formula (8.31) applies, except, of course, that $\bar{\xi}$ and $\bar{\xi}'$ refer to the extreme groups.

8.15   We may, in fact, go further (Theil, 1950) and consider the case in which the errors are not large enough to disturb the order of the members arranged according to one of the variables. Thus, assuming that the number of observations is even, say $2m$, there will correspond to any pair ξ_i, η_i the pair ξ_{m+i}, η_{m+i}, and we can form an estimator of α_1 from each of the m statistics

$$a_i = \frac{\eta_{m+i} - \eta_i}{\xi_{m+i} - \xi_i}. \tag{8.37}$$

We may choose any convenient average of these m estimates to estimate α_1.

Alternatively, we may construct similar estimates from all the $\frac{1}{2}n(n-1)$ pairs of sample numbers and choose some convenient average of the results.

Example 8.3

Reverting to the data of Example 8.2, again with the middle value omitted, we have four estimators of type (8.37). They are:

1.875, 1.903, 1.753, 1.887, mean 1.85, median (half-way between the middle values) 1.88, true value 2.

For the nine observations, the 36 possible values of the slope range from -2.529 to 5.111, mean 1.93, median 1.90.

In the ranking case it is also possible to see the rank numbers as instrumental variables. In Example 8.2 the nine values are already in order, so we number them 1 to 9 and have

$$\Sigma \, \zeta_i \eta_i = (1 \times 6\cdot9) + (2 \times 12\cdot5) + \ldots + (9 \times 39\cdot1) = 1267\cdot7$$
$$\Sigma \, \zeta_i \xi_i = (1 \times 1\cdot8) + (2 \times 4\cdot1) + \ldots + (9 \times 18\cdot9) = 549\cdot0.$$

We also have

$$\bar{\eta} = 23\cdot14, \qquad \bar{\xi} = 9\cdot57 \quad \text{and}$$
$$\Sigma \, \zeta_i = \tfrac{1}{2}n(n+1) = 45.$$

Thus, from the observed means the covariances are one ninth of

$$\Sigma \, \zeta\eta - \bar{\eta} \, \Sigma \, \zeta = 226\cdot40$$
$$\Sigma \, \zeta\xi - \bar{\xi} \, \Sigma \, \zeta = 118\cdot35$$

and the estimator of α_1 is

$$a_1 = 226\cdot40/118\cdot35 = 1\cdot91.$$

8.16 In order to simplify the expositions, we have, in the main, considered the relation between two variables Y and X. The methods can in many cases be extended to more general situations. For example, the controlled variable situation, as we have seen, extends to curvilinear data, and it is easy to see that any number of controlled variables can be handled in the same way. Likewise the method of groups can be employed on ϕ-variate material. For instance, if we can divide the data into k groups with respect to one variable which has the same observed order as for the unobserved variable which it represents, we can fit a hyperplane or even a curved surface (if k is large enough) through the centres of gravity of the groups. Or we can generalize the procedure of (8.36). All these methods depend on some assumption about the errors relative to the values X, which are often plausible enough in physical sciences but not always so in the behavioural sciences.

8.17 For instrumental variables and the grouping method it is even possible to put confidence regions on coefficients in certain simple cases. For example (Durbin, 1954), in the model of (8.7), $\eta - \alpha_0 - \alpha_1\xi = \epsilon - \alpha_1\delta$ and is uncorrelated with ζ. Then the observed correlation between ζ and $\eta - \alpha_0 - \alpha_1\xi$ is distributed as r from an uncorrelated population, and for normal variation this implies that

$$t^2 = \frac{(n-2)r^2}{1-r^2} \tag{8.38}$$

is distributed as the square of Student's t with $n - 2$ d.f. If we choose a confidence coefficient $1 - \theta$ and $t^2_{1-\theta}$ is such that

$$P(t^2 < t_{1-\theta}^2) = 1 - \theta,$$

then, since r^2 is a monotonic increasing function of t^2, being equal to $t^2/\{t^2 + (n-2)\}$, we have

$$P\left\{\frac{[\Sigma \zeta(\eta - \alpha_0 - \alpha_1 \xi)]^2}{\Sigma \zeta^2 \Sigma (\eta - \alpha_0 - \alpha_1 \xi)^2} < r_{1-\theta}^2\right\} = 1 - \theta. \tag{8.39}$$

The expression in curly brackets depends only on α_0, α_1 and the observables ξ, η, ζ and hence defines a quadratic confidence region for α_0 and α_1.

8.18 For the case of two equal groups we quote without proof a result of Wald (1944) that in the model of (8.7)

$$t = \frac{(\bar{\xi}' - \bar{\xi})(a_1 - \alpha_1)n^{\frac{1}{2}}}{2\{S_\eta^2 - 2\alpha_1 S_{\eta\xi} + \alpha_1^2 S_\xi^2\}^{\frac{1}{2}}} \tag{8.40}$$

is distributed as Student's t with $n - 2$ d.f. Here, with $n = 2m$,

$$S_\eta^2 = \frac{1}{n-2}\left\{\sum_{i=1}^{m} (\eta_i - \bar{\eta})^2 + \sum_{i=m+1}^{2m} (\bar{\eta}_i' - \bar{\eta}')'\right\} \tag{8.41}$$

and similar expressions for S_ξ^2 and $S_{\xi\eta}$. Equation (8.40) enables us to set confidence limits to α_1 regardless of α_0. It is also possible to set confidence regions for α_0 and α_1.

In the case of three groups, each one-third of the observations,

$$t = \left(\frac{n}{6}\right)^{\frac{1}{2}} \frac{(\bar{\xi}' - \bar{\xi})(a_1 - \alpha_1)}{S} \tag{8.42}$$

is distributed as t with $n - 3$ d.f. Here S^2 is given by

$$S^2 = S_\eta^2 - 2\alpha_1 S_{\xi\eta} + \alpha_1^2 S_\xi^2 \tag{8.43}$$

and the resemblance to (8.40) is clear.

8.19 The reader may feel that we have been making rather heavy weather of a very simple problem. Let us go back to a classical and more straightforward approach. We assume that ξ, η are jointly normally distributed (the same for all i). The model requires that $\xi_i = X_i + \delta_i$, and here we have to estimate not only the usual parameters α_0, α_1, var δ and var ϵ, four in number, but the n unknown values X_i. It is not surprising that we encounter difficulties, for we have more parameters than observations.

The likelihood function of δ and ϵ is proportional to

$$\frac{1}{\sigma_\delta^n \sigma_\epsilon^n} \exp\left[-\tfrac{1}{2}\Sigma \left(\frac{\delta^2}{\sigma_\delta^2} + \frac{\epsilon^2}{\sigma_\epsilon^2}\right)\right]$$

$$= \frac{1}{\sigma_\delta^n \sigma_\epsilon^n} \exp\left[-\frac{1}{2}\left\{\frac{1}{\sigma_\delta^2} \Sigma (\xi_i - X_i)^2 + \frac{1}{\sigma_\epsilon^2} \Sigma (\eta_i - \alpha_0 - \alpha_1 X_i)^2\right\}\right] \tag{8.44}$$

Thus

$$-\frac{2}{n}\log L + \text{constant} = \frac{S_1}{\sigma_\delta^2} + \log \sigma_\delta^2 + \frac{S_2}{\sigma_\epsilon^2} + \log \sigma_\epsilon^2, \qquad (8.45)$$

where

$$S_1 = \Sigma\,(\xi_i - X_i)^2, \quad S_2 = \Sigma\,(\eta_i - \alpha_0 - \alpha_1 X_i)^2. \qquad (8.46)$$

Now as $\sigma_\delta^2 \to 0$, $S_1/\sigma_\delta^2 + \log \sigma_\delta^2$ tends to $-\infty$ if $S_1 = 0$. The function on the right of (8.45) also tends to $-\infty$ if either of the constituent parts remains bounded while S_1 or S_2 is zero. Consider then

$$X_i = \xi_i, \quad \sigma_\delta^2 \to 0.$$

This will result in an infinite likelihood whatever the value of α_1. Hence the method of maximum likelihood does not work here, as Solari (1969) was the first to point out. Some further assumption is necessary to make a solution possible.

8.20 This need for a further assumption often seems strange to the user of statistical method, who has perhaps too much faith in its power to produce a simple and acceptable solution to any problem that can be posed simply. A geometrical illustration is therefore possibly useful.

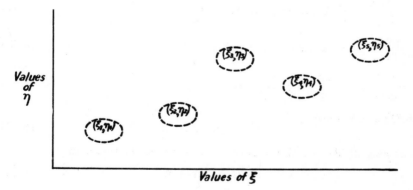

Fig. 8.1 Confidence regions for (X_i, Y_i) — see text

Consider the points (ξ_i, η_i) plotted as in Fig. 8.1. Any observed point (ξ_i, η_i) has emanated from a "true" point $(X_i, Y_i) = (\xi_i - \delta_i, \eta_i - \epsilon_i)$ whose situation is unknown. Since, in our model, δ_i and ϵ_i are independent normal variates, (ξ_i, η_i) is equiprobable on an ellipse centred at (X_i, Y_i) whose axes are parallel to the co-ordinate axes. Conversely, since the frequency function of ξ_i, η_i is symmetric in (ξ_i, η_i) and (X_i, Y_i), there is an elliptical confidence region for (X_i, Y_i) at any given probability level, centred at (ξ_i, η_i). These are the regions shown in Fig. 8.1. Heuristically, our problem of estimating α_1 may be conceived as that of finding a straight line to intersect as many as possible of these confidence regions. The difficulty is now plain to see: the problem as specified does not tell us what the lengths of the axes of the ellipses should be — these depend on the scale parameters σ_δ, σ_ϵ. It is clear that to make the problem definite we need only know the eccentricity of the ellipses, i.e. the ratio $\sigma_\delta/\sigma_\epsilon$.

8.21 Let us, then, suppose that $\sigma_\epsilon^2/\sigma_\delta^2 = \lambda$ is known. If we substitute for σ_δ^2 in (8.45), the latter becomes

$$-\frac{2}{n} \log L = -\log \lambda + 2 \log \sigma_\epsilon^2 + \frac{\lambda S_1 + S_2}{\sigma_\epsilon^2}. \tag{8.47}$$

(8.45) no longer makes this $\to -\infty$, for S_2 in the numerator of the last term remains positive. Moreover, it is obvious from the definitions of S_1 and S_2 that we cannot choose parameter values that will make $S_1 - S_2 = 0$. Thus (8.47) is bounded away from $-\infty$ and L itself from $+\infty$. We proceed to the ML solution through the likelihood equations. We find

$$\frac{\partial \log L}{\partial X_i} = \frac{1}{\sigma_\epsilon^2} [\lambda(\xi_i - X_i) + \alpha_1\{\eta_i - \alpha_0 - \alpha_1 X_i\}] = 0, \quad i = 1, 2, \ldots, n. \tag{8.48}$$

$$\frac{\partial \log L}{\partial \alpha_0} = \frac{1}{\sigma_\epsilon^2} \sum_i \{\eta_i - \alpha_0 - \alpha_1 X_i\} = 0 \tag{8.49}$$

$$\frac{\partial \log L}{\partial \alpha_1} = \frac{1}{\sigma_\epsilon^2} \sum_i X_i(\eta_i - \alpha_0 - \alpha_1 X_i) = 0 \tag{8.50}$$

$$\frac{\partial \log L}{\partial \sigma_\epsilon} = -\frac{2n}{\sigma_\epsilon^2} + \frac{n}{\sigma_\epsilon^3}(\lambda S_1 + S_2) = 0. \tag{8.51}$$

(8.51) at once gives

$$\sigma_\epsilon^2 = \tfrac{1}{2}(\lambda S_1 + S_2). \tag{8.52}$$

Summing (8.48) over all values of i, and using (8.49), we obtain

$$\sum_i (\xi_i - X_i) = 0 \tag{8.53}$$

and if we measure the ξ_i from their observed mean, we have from (8.53) the ML estimator of ΣX_i

$$(\Sigma X_i) = \Sigma \xi_i = 0. \tag{8.54}$$

Using (8.54) in (8.49), we obtain

$$\Sigma \eta_i = n\alpha_0$$

and if we measure the η_i also from their observed mean, this gives

$$\hat{\alpha}_0 = 0, \tag{8.55}$$

which reduces (8.52) to

$$\sigma_\epsilon^2 = \frac{1}{2n}\left\{\lambda \sum_i (\xi_i - X_i)^2 + \sum_i (\eta_i - \alpha_1 X_i)^2\right\}. \tag{8.56}$$

(8.48) now gives directly, using (8.55),

$$\lambda(\xi_i - X_i) + \hat{\alpha}_1(\eta_i - \hat{\alpha}_1 X_i) = 0$$

or

$$X_i = \frac{\lambda\xi_i + \hat{\alpha}_1\eta_i}{\lambda + \hat{\alpha}_1^2}. \tag{8.57}$$

Using (8.57) and (8.55) in (8.50), we find

$$\hat{\alpha}_1 = \frac{(\lambda + \hat{\alpha}_1^2)\{\lambda \sum \xi_i\eta_i + \hat{\alpha}_1 \sum \eta_i^2\}}{\lambda^2 \sum \xi_i^2 + \hat{\alpha}_1^2 \sum \eta_i^2 + 2\lambda\hat{\alpha}_1 \sum \xi_i\eta_i}$$

which simplifies to

$$\hat{\alpha}_1^2 \sum \xi_i\eta_i + \hat{\alpha}_1(\lambda \sum \xi_i^2 - \sum \eta_i^2) - \lambda \sum \xi_i\eta_i = 0 \tag{8.58}$$

or

$$\hat{\alpha}_1^2 + \hat{\alpha}_1 \frac{\lambda \sum \xi_i^2 - \sum \eta_i^2}{\sum \xi_i\eta_i} - \lambda = 0. \tag{8.59}$$

The positive root in $\hat{\alpha}_1$ must be chosen. For

$$\left.\begin{aligned}\hat{\alpha}_1 &= \frac{\sum \eta_i^2 - \lambda \sum \xi_i^2 \pm \{(\sum \eta_i^2 - \lambda \sum \xi_i^2)^2 + 4\lambda(\sum \xi_i\eta_i)^2\}^{\frac{1}{2}}}{2 \sum \xi_i\eta_i}\\ &= N/(2 \sum \xi_i\eta_i), \text{ say}\end{aligned}\right\} \tag{8.60}$$

and

$$\text{est var } X = \sum \xi_i\eta_i/\hat{\alpha}_1 = 2(\sum \xi_i\eta_i)^2/N > 0.$$

8.22   It is not by any means obvious that the estimator $\hat{\alpha}_1$ so derived is consistent, i.e. tends to the true value in large samples. We may, however, show that this is so. We have

$$\text{var } \xi = \text{var } X + \text{var } \epsilon/\lambda, \quad \text{var } \eta = \alpha_1^2 \text{ var } X + \lambda \text{ var } \delta \tag{8.61}$$
$$\text{cov } (\xi, \eta) = \alpha_1 \text{ var } X.$$

Substituting for the corresponding sample values in (8.59), we have

$$\left.\begin{aligned}\hat{\alpha}_1 &\to [\{\alpha_1^2 \text{ var } X + \lambda \text{ var } \delta - \lambda(\text{var } X + \text{var } \delta)\}\\ &\quad + \{\alpha_1^2 \text{ var } X + \lambda \text{ var } \delta - \lambda(\text{var } X + \text{var } \delta)\}^2\\ &\quad + 4\lambda(\alpha_1 \text{ var}^2 X)^{\frac{1}{2}}]/2\alpha_1 \text{ var } X.\\ &= \alpha_1.\end{aligned}\right\} \tag{8.62}$$

However, $\hat{\sigma}_\epsilon^2$ is not a consistent estimator of σ_ϵ^2 as Lindley (1947) showed. In fact, from (8.56).

$$\hat{\sigma}_\epsilon^2 = \frac{\lambda}{2(\lambda + \hat{\alpha}_1^2)} \frac{1}{n} \Sigma (\eta_i - \hat{\alpha}_1 \xi_i)^2$$

$$= \frac{\lambda}{2(\lambda + \hat{\alpha}_1^2)} \{\text{var } \eta - 2\hat{\alpha}_1 \text{ cov } (\xi, \eta) + \hat{\alpha}_1^2 \text{ var } \xi\} \qquad (8.63)$$

and then from (8.61) and (8.62),

$$\hat{\sigma}_\epsilon^2 \rightarrow \frac{\lambda}{2(\lambda + \alpha_1^2)} \left\{\alpha_1^2 \text{ var } X + \sigma_\epsilon^2 + 2\alpha_1^2 \text{ var } X + \alpha_1^2 \left(\text{var } X + \frac{\sigma_\epsilon^2}{\lambda}\right)\right\} \qquad (8.64)$$

$$= \tfrac{1}{2}\sigma_\epsilon^2.$$

It appears that this inconsistency arises from the bias which is known to exist for quadratic maximum-likelihood estimators. (For example, $\Sigma (x - \bar{x})^2/n$ is a biassed estimator of parent variance.) In this particular case the adjustment to secure consistency is plain, but the case warns us against an over-ready acceptance of estimators in this type of situation until consistency has been proved.

Example 8.4

For the data of Example 8.2 we have $\lambda = 1$, $n = 9$ and

$$\Sigma \xi = 56\cdot1, \quad \Sigma \eta = 208\cdot3, \quad \bar{\xi} = 9\cdot57, \quad \bar{\eta} = 23\cdot14$$

$$n \text{ var } \xi = 238, \quad n \text{ var } \eta = 906, \quad n \text{ cov } (\xi, \eta) = 451$$

whence
$$\hat{\alpha}_1 = [(906 - 238) + \{(906 - 238)^2 + 4(451)^2\}^{\frac{1}{2}}]/(2 \times 451)$$

$$= 1\cdot99, \text{ a very close estimate.}$$

8.23   The geometrical interpretation of the line $Y = \alpha_0 + \alpha_1 X$ is particularly interesting. In fact, looking to the minimization of (8.44), we see that if $\sigma_\epsilon^2 = \sigma_\delta^2$ (or if we rescale so that this is so) we are minimizing the sum of squares of distances from the points to the fitted line. In short, if we know the ratio var ϵ/var δ and re-scale accordingly, the line of closest fit is the first eigenvector. We might, in fact, have taken the minimization of sums of squares of distance as a heuristic criterion of fit.

It is intuitively plausible, and can easily be proved, that the line so determined lies between the two regression lines. In fact, from (8.60) we see that if $\lambda = 0$, $\hat{\alpha}_1$ becomes $\Sigma \eta^2/\Sigma \xi\eta$, the slope of the regression of ξ on η; and if λ is $+\infty$, $\hat{\alpha}_1$ becomes $\Sigma \xi\eta/\Sigma \xi^2$, the slope of the regression of η on ξ.

8.24   From the same viewpoint, if we are estimating the constants in the general functional relationship

$$\sum_{j=1}^{k} \alpha_j X_j = \text{constant}, \qquad (8.65)$$

we require the proportionality of the error variances. If var $\delta_i = k\mu_i$, where k is a constant (perhaps unknown) and μ_i is known, we can rescale to unit variances for all ξ^2, and the hyperplane (8.65) corresponds to the first eigenvector.

8.25 It is natural to consider the extension to curvilinear relationship of the concept of minimal average distance or distance squared. Mathematically the problem then becomes much more complex than in the linear case, because the perpendiculars from the observed points to the fitted curve depend on the local shape of the curve and are not all parallel. One approach is to minimize the sum of squares parallel to the x-axis plus the sum parallel to the y-axis *weighted by some quantity* chosen at convenience. Another, more popular in the computer age, is to proceed by iteration, estimating the curve in a preliminary way in terms of parameters, finding the sum of squares of distances, revising the parameters of the curve, and so on; one of the possible methods is that generally known as Newton—Raphson. The curves fitted in this way have an empirically good fit, but they are free from the theoretical difficulties which we discussed earlier in appearance only. In fact, whenever we assign distances or drop perpendiculars, we are dependent on the relative scales of the variables. The difficulties are enhanced if we extend our requirements to fitting surfaces in three or more dimensions.

8.26 The problems of specification which we encountered in discussing factor analysis and some of the problems of fitting functional relationships are particular cases of what, in econometrics, have become known as problems of *identifiability*. It is possible to write down models which are unidentifiable in the sense that some of the parameters are theoretically inestimable, however much information is available. This is an unacceptable situation in science, whatever may be the situation in metaphysics. We are therefore faced with the necessity (1) of recognizing unidentifiable models when they occur – not always a simple procedure, and (2) of modifying the model to make it identifiable. There is a surprisingly wide range of resources for this purpose, but each situation has to be considered on its merits. In econometrics one may sometimes add further variables to one equation, add further equations, impose conditions on the parameters from extraneous considerations, or use cross-section data to assign values to some of them. In physics we may replicate experiments to gain some idea of relative experimental errors. In this chapter we have resolved the identifiability problem by imposing further constraints on the model; by assuming that we know something about relative errors, or, as in the grouping and ranking methods, assuming that the errors are small enough not to disturb the observed order of some of the variables. The chapter may be regarded as an illustration of the importance of having regard to the underlying model which, in multivariate analysis, is often vaguely specified and may, in careless hands, present the analytical statistician with an impossible task.

NOTES

(1) A more detailed treatment of the subject, including problems of structural relationship (i.e. relations among random variables as distinct from functional relationship among mathematical variables) is given in Kendall and Stuart, vol. 2, chapter 29.

(2) The identifiability problem in econometrics is usually bound up with time-series effects. Reference may be made to Fisher (1957) or Fisk (1967), for comprehensive accounts.

9
Tests of hypotheses

9.1 In univariate theory there are several criteria by which we compare two or more hypotheses. One, of very general use, is based on the ratio of likelihoods. Suppose that the likelihood of the sample on hypothesis H_0 is L_0, and that of H_1 is L_1. The ratio L_0/L_1 gives us some measure of the "closeness" of the two hypotheses. If they are identical the ratio is unity. As they diverge, the ratio diminishes to zero or increases. (It cannot, of course, be negative.) In fact, we shall find it more convenient to use the monotonically related function $\log L_0/L_1$ and to consider H_0 as more general than H_1, which concerns parameters forming a subset of the wider class in H_0. Then L_1 will be greater than L_0, and the ratio L_0/L_1 varies between 0 and 1. The quantity $\log(L_0/L_1)$ varies from 0 to ∞.

9.2 The use of the criterion in practice requires a specification of the form of the parent population, and we shall, throughout this chapter, assume it to be multivariate normal. We do not, however, know the parameters − if we did, there would be no problems of estimation or hypothesis testing. We shall, therefore, estimate those parameters from the sample and substitute in the likelihood function, so obtaining an estimate of the likelihood. The ratio of two such estimates then depends solely on the sample observations and is a statistic with a sampling distribution. This distribution, to which, fortunately, we shall be able to find simple approximations for large samples, provides the basis of a test of the hypothesis, or rather, of a comparison between two hypotheses. (No hypothesis is tested *in vacuo*; only by comparison with at least one alternative.)

9.3 In the first instance we consider the estimation of means and covariances in multivariate normal samples, and we shall make use of the principle of maximum likelihood. One property of maximization procedures is worth pointing out. If we are maximizing a function, say $f(\theta_1, \theta_2)$ for both parameters, we may solve the simultaneous equations:

$$\partial f/\partial\theta_1 = 0, \quad \partial f/\partial\theta_2 = 0. \tag{9.1}$$

125

Equivalently we may solve $\partial f / \partial \theta_1$ for θ_1, substitute in $f(\theta_1, \theta_2)$ and solve the resultant $\partial f / \partial \theta_2$ for θ_2. This gives the same result.

9.4   Consider then k multivariate normal populations with means typified by μ_{jt} $(j = 1, 2, \ldots, p; t = 1, 2, \ldots, k)$ and dispersions γ_{jlt}, or equivalently $\sigma_{jt}\sigma_{lt}\rho_{jlt}$ $(j, l = 1, 2, \ldots, p)$. Let there be a sample of n_t from the tth population. If α_{jlt} is inverse to γ_{jlt}, the likelihood function of all samples together is

$$\left\{\prod_{t=1}^{k} \frac{|\alpha_t|^{\frac{1}{2}n_t}}{(2\pi)^{\frac{1}{2}pn_t}}\right\} \exp\left\{-\tfrac{1}{2} \sum_{t=1}^{k} \underset{n_t}{S} \sum_{j,l=1}^{p} \alpha_{jlt}(x_{jt} - \mu_{jt})(x_{lt} - \mu_{lt})\right\} . \quad (9.2)$$

The symbol S means exactly the same as the summatory sign Σ. We write it so merely to remind us that it relates to summation over the sample.

The logarithm of this expression is the sum of k independent logarithms each of which can be maximized by itself. Differentiation with respect to μ_{jt} gives us for the ML estimators of means

$$\frac{\partial}{\partial \mu_{jt}} \log L = S \Sigma \alpha_{jlt}(x_{jt} - \hat{\mu}_{jt}) = 0, \quad (9.3)$$

equivalent to

$$\Sigma \alpha_{jlt}(\bar{x}_{jt} - \hat{\mu}_{jt}) = 0. \quad (9.4)$$

Since α is not degenerate, the p equations of this type (one for each j) are equivalent to

$$\hat{\mu}_{jt} = \bar{x}_{jt}, \quad (9.5)$$

that is, the mean of the sample is an ML estimator of the corresponding parent mean.

9.5   Likewise, for the parameter α we have

$$\tfrac{1}{2}n_t \frac{1}{|\hat{\alpha}_t|} \frac{\partial |\hat{\alpha}_t|}{\partial \hat{\alpha}_{jlt}} - \tfrac{1}{2}S(x_{jt} - \hat{\mu}_{jt})(x_{lt} - \hat{\mu}_{lt}) = 0. \quad (9.6)$$

Now the co-factor of α_{jlt} in $|\alpha_t|$ divided by $|\alpha_t|$ is the inverse γ_{lt}, so (9.6) reduces to

$$\hat{\gamma}_{jlt} = c_{jlt}, \quad (9.7)$$

where c_{jlt} is the covariance defined with divisor n_t (not $n_t - 1$). Thus the sample covariance, so defined, is the ML estimator of the corresponding parent covariance. It will follow that the sample correlations are also ML estimators of the parent correlations.

9.6   If we substitute these estimates in the likelihood of (9.2) a remarkable simplification appears. In fact, the exponent becomes the product of terms like

$$-\tfrac{1}{2}n_t \sum_{j,l=1}^{p} \hat{\alpha}_{jlt}\hat{\gamma}_{jlt} = -\tfrac{1}{2}n_t.$$

Thus the whole likelihood reduces to

$$L = \frac{e^{-\frac{1}{2}n}}{(2\pi)^{\frac{1}{2}np}} \prod_{t=1}^{k} \frac{1}{|c_{jlt}|^{\frac{1}{2}n_t}},$$ (9.8)

where n is the total sample size $\sum_{t=1}^{k} n_t$. Here, as elsewhere, we are led naturally to determinants of the covariance type.

9.7   Now suppose that we wish to consider the hypothesis that the k populations are identical in means and covariances, as against the situation we have just examined in which they are all different. The k contributions in (9.2) then are melded together to give a likelihood L_0, say:

$$L_0 = \frac{|\alpha|^{\frac{1}{2}n}}{(2\pi)^{\frac{1}{2}np}} \exp\left\{-\tfrac{1}{2}S \sum_{j,l=1}^{p} \alpha_{jl}(x_j - \mu_j)(x_l - \mu_l)\right\}.$$ (9.9)

In the same manner as before, we find that the means are estimated by the means of the pooled samples, and so for the covariances. The estimated likelihood then becomes

$$L_0 = \frac{e^{-\frac{1}{2}n}}{(2\pi)^{\frac{1}{2}np}} \cdot \frac{1}{|c_{jl}|^{\frac{1}{2}n}}.$$ (9.10)

Thus, to test the hypothesis, say H, that the k samples are homogeneous (come from identical populations) as against the alternative that they come from different populations, we test the ratio of (9.10) divided by (9.8), namely

$$l_H = \frac{\prod |c_{jlt}|^{\frac{1}{2}n_t}}{|c_{jl}|^{\frac{1}{2}n}} = \prod_{t=1}^{k} \left\{\frac{|c_{jlt}|}{|c_{jl}|}\right\}^{\frac{1}{2}n_t}.$$ (9.11)

The maximization at (9.8) has more free parameters than that at (9.10), and the corresponding estimated likelihood must accordingly be greater. Thus l_H of (9.11) varies from 0 to 1. If it is close to unity, we accept the hypothesis of homogeneity; the further it diminishes towards zero the more we incline to the alternative hypothesis of heterogeneity.

9.8   We shall, in fact, following the usual procedure in analysis of variance, consider three hypotheses:

H (the one we have just discussed) that the populations are identical, having the same means and dispersions;

H_1 we admit that the means may be different, and our hypothesis is whether the populations have the same dispersions;

H_2 it being known that they have the same dispersions, that they differ in the means.

To test H_1, we have to consider the likelihood (9.2) when the α_{jlt} are the same for all t but the means are not. As before, the means are estimated by the sample means. The common dispersion, which we write as γ_{jla}, is now an average of the individual dispersions about the different means, and is estimated by

$$c_{jla} = \frac{1}{n} \sum_{t=1}^{k} \sum_{u=1}^{n_t} (x_{jut} - \bar{x}_{jt})(x_{lut} - \bar{x}_{lt}). \qquad (9.12)$$

The criterion to test the hypothesis H_1 against the alternative of heterogenity in means and covariances then becomes

$$l_{H_1} = \prod_{t=1}^{k} \left\{ \frac{|c_{jlt}|}{|c_{jla}|} \right\}^{\frac{1}{2}n_t}, \qquad (9.13)$$

again a ratio of covariance type determinants.

9.9 In a similar way, if we assume that all dispersions are equal but test H_2, that their means are the same (as against the alternative that the dispersions are equal and their means differ), the criterion becomes

$$l_{H_2} = \prod_{t=1}^{k} \left\{ \frac{|c_{jla}|}{|c_{jl}|} \right\}^{\frac{1}{2}n_t} = \left\{ \frac{|c_{jla}|}{|c_{jl}|} \right\}^{\frac{1}{2}n}. \qquad (9.14)$$

We note that $l_H = l_{H_1} l_{H_2}$. We may also observe that, by a splitting of quadratic expressions familiar in variance analysis,

$$c_{jl} = \frac{1}{n} \underset{n}{S} (x_{jt} - \bar{x}_j)(x_{lt} - \bar{x}_l)$$

$$= \frac{1}{n} \sum_{t=1}^{k} S(x_{jt} - \bar{x}_{jt})(x_{lt} - \bar{x}_{lt}) + \frac{1}{n} \sum_{t=1}^{k} n_t(x_{jt} - \bar{x}_j)(x_{lt} - \bar{x}_l)$$

$$= c_{jla} + c_{jlm}, \qquad (9.15)$$

where c_{jlm} is a weighted covariance of sample means about the means of pooled samples.

9.10 It is interesting to consider what happens to these criteria in the univariate case when $p = 1$. We find that l_H itself is simply unity. The others are less nugatory. It is easily seen that, apart from the power $\frac{1}{2}n$, l_{H_1} becomes effectively the ratio of the sum of squares within samples to the sum of squares between samples; and l_{H_2} becomes the ratio of the sum of squares between samples to the sum of squares of pooled samples. In ordinary ANOVA we customarily test l_{H_2}, the sum of squares between samples, against the residual variance within samples; here in the multivariate case the ratio is between within-sample and total, not within-sample and total-less-within-sample.

9.11   To complete the tests we have to find the sampling distribution of the various criteria. The exact distribution theory in such cases is complicated, and the situation resembles the one we noticed in Chapter 6. The moments of the criteria can be derived but are fairly complicated (for details see Kendall and Stuart, vol. 3). However, for all practical purposes we can appeal to an asymptotic result, to the general effect that the logarithms of likelihood ratios based on ML estimators (or simple multiples of them) can be tested in the χ^2 distribution. We will sketch the proof of this important result.

(1)   In the first place, it is known that if C is a p-variate quadratic form and $k \exp(-\frac{1}{2}Q)$ is multivariate normal, then Q itself is distributed as χ^2 with p degrees of freedom. This follows easily from the fact that Q can be transformed through its eigenvectors or otherwise to the sum of p squares of normal variables.

(2)   Secondly, if a set of parameters $\boldsymbol{\theta}$ are estimated by ML methods and $\hat{\boldsymbol{\theta}}$ are the estimators, then asymptotically

$$-\tfrac{1}{2}(\hat{\boldsymbol{\theta}} - \boldsymbol{\theta})'\mathbf{V}^{-1}(\hat{\boldsymbol{\theta}} - \boldsymbol{\theta}) \qquad (9.16)$$

is multivariate normal with covariance matrix $\mathbf{V}$, which is given by

$$\mathbf{V}^{-1} = n\left[-\,\mathrm{E}\,\frac{\partial^2 \log L}{\partial\theta_r\,\partial\theta_s}\right]. \qquad (9.17)$$

This is the generalization of the univariate result that for a single parameter θ,

$$\mathrm{var}\,\hat{\theta} = -1/n\,\mathrm{E}\,\frac{\partial^2 \log L}{\partial\theta^2}. \qquad (9.18)$$

(3)   Now consider the case in which there are $r + s$ parameters and we test $H_0: \boldsymbol{\theta}_r = \boldsymbol{\theta}_0$ (θ_s free), against H_1 ($\boldsymbol{\theta}_r \neq \boldsymbol{\theta}_0$) ($\theta_r$ and θ_s free). We find for H_1 the likelihood when all parameters are estimated, giving, say, an unconditional maximum $L(\hat{\theta}_r, \hat{\theta}_s)$. For H_0 we find the conditional maximum $L(\theta_r, \hat{\hat{\theta}}_s)$, where $\hat{\hat{\theta}}_s$ may be different from $\hat{\theta}_s$. Consider the ratio

$$l = \frac{L(\theta_r, \hat{\hat{\theta}}_s)}{L(\hat{\theta}_r, \hat{\theta}_s)}. \qquad (9.19)$$

For large samples the denominator tends to multivariate normality, but then $\hat{\theta}_r, \hat{\theta}_s$ tend to θ_r, θ_s and L is unity, except for a constant. The numerator tends to $-\frac{1}{2}$ times the exponential of a quantity proportional to a quadratic function in the x's, subject to s linear constraints. This is transformable to a χ^2 with r degrees of freedom. Thus, $-2\log l$ is distributed as χ^2 with r d.f.

9.12 In spite of the initial complexity, then, our l-ratios, or rather their logarithms, can be tested in χ^2. The number of degrees of freedom is the number of parameters estimated in the denominator less the number estimated in the numerator.

Example 9.1 (Pearson and Wilks, 1933)

Five samples are available, each of twelve members, of aluminium die-castings ($k = 5$, $n_t = 12$ for all t, $n = 60$). On each specimen two measurements are taken: tensile strength (1000 lb per square inch) which we call x, and hardness (Rockwell's E) which we call y. The data may be summarized as follows:

Reference number of sample t	x		y		Correlation coefficient
	Mean	Standard deviation	Mean	Standard deviation	
1	33·399	2·565	68·49	10·19	0·683
2	28·216	4·318	68·02	14·49	0·876
3	30·313	2·188	66·57	10·17	0·714
4	33·150	3·954	76·12	11·18	0·715
5	34·269	2·715	69·92	9·88	0·805

We are interested in the homogeneity of these data.

We first of all test H_1, that the data show no significant difference in dispersions. We have the following results:

t	Sums of squares about means		Sums of products about means	Covariance determinants	$\log_{10}$ of cov. determinants
	x	y			
1	78·948	1247·18	214·18	365·204	2·562 54
2	223·695	2519·31	657·62	910·401	2·959 23
3	57·448	1241·78	190·63	243·029	2·385 66
4	187·618	1473·44	375·91	938·451	2·972 41
5	88·456	1171·73	259·18	253·281	2·403 60
Totals	636·165	7653·44	1697·52	—	13·283 44

For the pooled variances and covariances about respective means we have

$$c_{11a} = 636\cdot165/60 = \quad 10\cdot6028$$
$$c_{22a} = \quad\quad\quad\quad\quad\quad 127\cdot5573$$
$$c_{12a} = \quad\quad\quad\quad\quad\quad 28\cdot2920$$
$$|c_{jla}| = \quad\quad\quad\quad\quad\quad 552\cdot0180$$

The criterion we require is that of (9.13). In this case each $n_1 = 12$, so

$$-2 \log_e l = -2 \cdot \tfrac{12}{2} \sum_s \{\log_{10} |c_{jlt}| - \log_{10} |c_{jla}|\} \log_e 10$$

$$= -12\{13 \cdot 283\,44 - 13 \cdot 709\,86\} \log_e 10 = 11 \cdot 78.$$

The degrees of freedom number 12. For the unrestricted case there are five parameters per sample — two means and three covariances — 25 in all. For the restricted sample there are 10 means and three covariances, 13 in all. The difference is 12.

The value of χ^2 is so close to 12 that a test is unnecessary. We therefore accept the hypothesis of homogeneity in the covariances. We can then proceed to H_2, that the means are homogeneous. For this we require the covariance determinant of all data pooled together. For this, the sum of squares of x is 942·254, that of y is 8316·19, and the sum of xy is 1912·38. $|c_{ij}|$ is then 1160·77. We find also $|c_{ijm}| = 43 \cdot 526$, $|c_{ija}| = 552 \cdot 018$. The criterion (9.14) gives then $-12 \sum_s \log_e(552 \cdot 018 / 1160 \cdot 77) = 44 \cdot 59$. The number of degrees of freedom is now $13 - 5 = 8$. The observed value of χ^2 is highly significant. We are led to suppose that the samples differ in their means but not in their variances. The question then arises whether it is the mean of x or the mean of y, or both, which give rise to this heterogeneity. Strictly speaking, owing to the correlation between the x and y we cannot conduct independent tests on them, but it is worth while seeing whether separate analyses of variance on x and y are suggestive. We can treat each as an ordinary ANOVA with a one-way classification and we obtain the following:

	Estimates of variance		d.f.
	x	y	
Between samples	76·522	165·69	4
Within samples	11·566	139·15	55

An ordinary F-test indicates that at the 1 percent point the differences between tensile strength, but not those between hardness, are responsible for the heterogeneity.

Example 9.2 (Pearson and Wilks, 1933)

Measurements were available on the lengths and breadths of 600 skulls, 20 from each of 30 races. (The details are given in the paper under reference.) That there would be some variation between races was to be expected, but it was of interest to see whether it extended to dispersions. The hypothesis H_1 was tested. The authors found

$$|c_{ija}| = 656 \cdot 369 \quad 1/k \sum_{t=1}^{k} \log\{|c_{ijt}|\} = 2 \cdot 644\,429$$

$$\log|c_{ija}| = 2 \cdot 817\,148$$

$$\text{Difference} = \bar{1} \cdot 827\,281$$

To base e this logarithm is $-0 \cdot 3977$. The simple test gives $600(0 \cdot 3977)$ $= 238 \cdot 6$ distributed as χ^2 with $150 - 63 = 87$ d.f. This is highly significant.

There is therefore lack of uniformity in the dispersions. We now proceed to consider them individually.

The homogeneity of a set of variances in the univariate case can be tested by known methods which are, in fact, simpler forms of the likelihood criteria that we are employing here. It is found that the variances of both x and y are significantly heterogeneous. Finally we may consider the correlations. A familiar result in elementary theory states that the transformation

$$z = \tanh^{-1} r$$

reduces z to approximate normality with variance $n - 3$. We therefore test

$$\chi^2 = \sum_{t=1}^{30} (n_t - 3)(z_t - \bar{z})^2$$

with 29 degrees of freedom. In the present instance this gave a χ^2 value of $96 \cdot 01$, which is very significant.

The general conclusion is that there is racial heterogeneity in every respect, in the dispersion of head length and width about their means and in the correlation between the two measurements.

9.13   The question arises as to how we should test the differences between means if the dispersions are found to be heterogeneous. Since H_1 is untenable, we should not test H_2. There is no clear answer to this question. We may be willing to assume that the heterogeneity of covariances is not serious enough to invalidate the test of H_2, which is somewhat hazardous; or we may be prepared to test the means of the variables separately as a set of one-way ANOVA, notwithstanding what has been said in the foregoing about correlation between tests. It seems that a certain amount of subjective judgement is required in such cases. In ordinary ANOVA, where we sometimes test H_2 without a prior test of H_1, we are apparently saved from error by the fact that H_2 is not very sensitive to small departures from equality in the variances within classes, especially if the class numbers are about the same. In MANOVA there are indications that the test of H_2 is sensitive to differences between parent correlations, though not, perhaps, to differences between variances. It is desirable to inspect the correlation matrices of the individual groups to see how similar they are before assuming that the test of H_2 is robust.

Example 9.3 (Wilks, 1946)

The kind of technique used above to test the differences between populations may also be used to test the internal structure of a single population.

A test involving 60 items was given to 100 subjects and the scores noted. The test items were divided into three groups of 20 on the basis of external criteria. The question is whether these three groups are similar. We may regard the scores as 100 observations on a trivariate normal population, and the questions are whether the variances of the three variables are similar, and whether the covariances between variables are similar. We consider the general problem in p dimensions.

The hypothesis is that all means are equal, $\mu_i = \mu$, and that the covariance matrix is of the form

$$
\begin{bmatrix}
\sigma^2 & \rho\sigma^2 & \ldots & \rho\sigma^2 \\
\rho\sigma^2 & \sigma^2 & \ldots & \rho\sigma^2 \\
\cdot & \cdot & \cdots & \cdot \\
\rho\sigma^2 & \rho\sigma^2 & \ldots & \sigma^2
\end{bmatrix}
\tag{9.20}
$$

For unconstrained estimation of the parameters in a p-variate population we have, as at (9.10),

$$
L = \frac{e^{-\frac{1}{2}np}}{(2\pi)^{\frac{1}{2}np}} \cdot \frac{1}{|c_{jl}|^{\frac{1}{2}n}} .
\tag{9.21}
$$

We require to invert the matrix (9.20), obtaining

$$
\boldsymbol{\alpha} =
\begin{bmatrix}
A & B & \ldots & B \\
B & A & \ldots & B \\
\cdot & \cdot & \cdots & \cdot \\
B & B & \ldots & A
\end{bmatrix}
\tag{9.22}
$$

where

$$
A = \frac{1 + (p-2)\rho}{\sigma^2(1-\rho)\{1 + (p-1)\rho\}}
\tag{9.23}
$$

$$
B = \frac{-\rho}{\sigma^2(1-\rho)\{1 + (p-1)\rho\}} .
\tag{9.24}
$$

Thus the likelihood under the constrained hypothesis is

$$
\frac{(A-B)^{p-1}\{A + (p-1)B\}^{\frac{1}{2}n}}{(2\pi)^{\frac{1}{2}np}}
$$
$$
\times \exp\left[-\tfrac{1}{2}\{AS \sum_i (x_{it} - \mu)^2 + BS \sum (x_{it} - \mu)(x_{jt} - \mu)\}\right].
\tag{9.25}
$$

For the ML estimators there result

$$\hat{\mu}_i = \bar{x}_i$$

$$\hat{A} = \frac{1 + (p - 2)r_0}{s_0^2(1 - r_0)\{1 + (p - 1)r_0\}} \tag{9.26}$$

$$\hat{B} = \frac{-r_0}{s_0^2(1 - r_0)\{1 + (p - 1)r_0\}}, \tag{9.27}$$

where we define

$$c_{ij0} = c_{ij} + (\bar{x}_i - \bar{x})(\bar{x}_j - \bar{x}) \tag{9.28}$$

$$s_0^2 = \frac{1}{p} \sum_{i=1}^{p} c_{ii0} \tag{9.29}$$

$$r_0 = \sum_{i \neq j}^{p} c_{ij0} \Big/ \Big\{(p - 1) \sum_{i=1}^{p} c_{ii0}\Big\}. \tag{9.30}$$

The likelihood (9.25) then reduces to

$$\frac{e^{-\frac{1}{2}pn}}{(s_0^2)^p(1 - r_0)^{p-1}\{1 + (p - 1)r_0\}^{\frac{1}{2}n}(2\pi)^{\frac{1}{2}pn}}, \tag{9.31}$$

and the likelihood ratio is the $(\frac{1}{2}n)$th power of

$$\frac{|c_{ji}|}{(s_0^2)^p(1 - r_0)^{p-1}\{1 + (p - 1)r_0\}}. \tag{9.32}$$

This can be tested in χ^2 with $\frac{1}{2}p(p + 3) - 3$ degrees of freedom.

In a similar way, we find for hypothesis H_1 (that dispersions are equal regardless of means) the $(\frac{1}{2}n)$th power of

$$\frac{|c_{ij}|}{(s^2)^p(1 - r)^{p-1}\{1 + (p - 1)r\}}, \tag{9.33}$$

where s^2 and r refer to all variables pooled, with $\frac{1}{2}p(p + 3) - (p + 2)$ d.f.; and for hypothesis H_2 (given covariances, that means are equal), the $\frac{1}{2}n(p - 1)$th power of

$$\frac{s^2(1 - r)}{s^2(1 - r) + \sum_{i=1}^{p} (\bar{x}_i - \bar{x})^2/(p - 1)}, \tag{9.34}$$

with $\frac{1}{2}p(p + 3) - \frac{1}{2}p(p - 1) - 1 = p - 1$ d.f.

In the particular case under consideration,

$$\bar{x}_1 = 10\cdot9900 \qquad c_{11} = 16\cdot8451 \qquad c_{12} = 13\cdot5493$$
$$\bar{x}_2 = 10\cdot9300 \qquad c_{22} = 18\cdot1099 \qquad c_{13} = 14\cdot5826$$
$$\bar{x}_3 = 11\cdot2610 \qquad c_{33} = 17\cdot7134 \qquad c_{23} = 13\cdot8056$$
$$s^2 = 17\cdot5558 \qquad s_0^2 = 17\cdot5764$$
$$r = 0\cdot7963 \qquad r_0 = 0\cdot7963$$
$$|c_{ij}| = 545\cdot5308$$

The data look homogeneous, but we may apply a systematic test. For hypothesis H it is found that the criterion of (9.32) is the $(\frac{1}{2}n)$th power of 0·9209. The quantity $-2\log_e l$ is then $-100\log_e(0\cdot9209) = 8\cdot22$ with 6 degrees of freedom. This is not significant at the 20 percent level. We accept the hypothesis of homogeneity.

Had it been otherwise, we might have gone on to test H_1. The criterion of (9.33) becomes the $(\frac{1}{2}n)$th power of 0·9370. The criterion is then 6·51 with 4 d.f. Again the value is not significant and we accept the hypothesis of homogeneity in the covariances.

Finally, for the criterion of (9.34) we have the $\frac{1}{2}n(p-1)$th power of 0·9914. The criterion is 1·74 with 2 d.f.

We conclude that the three sub-tests are similar in respect of means and dispersions.

9.14   Just as we have an extension of ANOVA to the analysis of covariance, so we may extend MANOVA to testing the homogeneity of p-variate complexes after their dependence on other variables has been removed by regression analysis. In the ANOVA case, for example, suppose we observe values of y in k classes and are interested in the homogeneity of y over the classes. We may suspect that, apart from class effects, y is dependent on variables $x_1, \ldots, x_p$. We eliminate the effect of the x's by regressing y on them and considering the residuals $y - \Sigma b_j x_j$, the sums of squares of which are easily calculable. If these residuals vary significantly from class to class, there is evidence of a class effect distinct from the dependence of y on the x's. The same technique can be carried out in MANOVA. If a set of y's are dependent on a set of x's, we "eliminate" the x effect by regression and study the homogeneity of the residuals by tests based on the ratio of dispersion determinants, with appropriate allowance for degrees of freedom. The following two examples illustrate the method and type of argument.

Example 9.4 (Bartlett, 1947)

In an experiment to examine the effect of fertilizers on grain, eight treatments were applied in each of eight randomized blocks; and on each plot two observations were made, the yield of straw (x_1) and the yield of grain (x_2). The following was obtained:

	d.f.	s.s. (x_1^2)	s.p. $(x_1 x_2)$	s.s. (x_2^2)	
Blocks	7	86 045·8	56 073·6	75 841·5	
Treatments	7	12 496·8	−6 786·6	32 985·0	
Residual	49	136 972·6	58 549·0	71 496·1	
Total	63	235 515·2	107 836·0	180 322·6	(9.35)

We are not interested in block differences and extract them from the variation. This gives us the middle two lines of the table and a new total:

Total
(excluding
blocks):   56   149 469·4   51 762·4   104 481·1   (9.36)

A test of homogeneity of means is accordingly given by

$$\begin{vmatrix} 136\,972\cdot6 & 58\,549\cdot0 \\ 58\,549\cdot0 & 71\,496\cdot1 \end{vmatrix} \bigg/ \begin{vmatrix} 149\,469\cdot4 & 51\,762\cdot4 \\ 51\,762\cdot4 & 104\,481\cdot1 \end{vmatrix} = 0\cdot4920$$

and $56 \log_e 0\cdot4920 = 39\cdot7$ with $16 - 2 = 14$ d.f. The result is highly significant.

We can proceed to further analysis in several ways. First, in Example 9.1 we can examine the significance of the variates separately by an ordinary F-ratio. This gives

| | Estimated mean squares | | d.f. |
	x_1	x_2	
Treatments	1785	4712	7
Residual	2795	1459	49	(9.37)

The first ratio, 0·639 for 7 and 49 d.f., is fortunately not significant. (We should have been hard put to it to explain a significance if it had appeared, for the residual mean square is larger than the treatment mean square.) The second ratio is 3·230 for 7 and 49 d.f. which is significant at the 1 percent point. At first sight it would seem that the treatments are affecting grain yield but not straw yield. The correlation "between treatments" is about $-0\cdot3$ and that "between residuals" is about $+0\cdot6$, but the former is not significant.

We could also perform a covariance analysis to see if the treatment on straw was influenced by (and perhaps masked by) the concomitant variate grain. Our table is

	d.f.	s.s. (x_1^2)	s.p. $(x_1 x_2)$	s.s. (x_2^2)
Treatment	7	12 496·8	$-6\,786\cdot6$	32 985·0
Residual	49	136 972·6	58 549·0	71 496·1
Total	56	149 469·4	51 762·4	104 481·1	(9.38)

Considering the regression of x_1 on x_2, we have for the coefficient in the regression equation (calculated from the residual items)

$$b, \text{say}, = \frac{S(x_1 x_2)}{S(x_2^2)} = 0\cdot818\,911\,8.$$

The residual total s.s. (x_1^2) is then

$$136\,972\cdot6 - 58\,549\cdot0\,b = 89\cdot026\,1.$$

Similarly for the "total" sums of squares and products:

$$b', \text{say} = \frac{51\,762\cdot4}{104\,481\cdot1} = 0\cdot495\,423\,6,$$

and the residual sum of squares is

$$149\,469\cdot4 - 51\,762\cdot4\,b' = 128\,825\cdot1.$$

We then construct the table for residual x_1 effects after the extraction of x_2, obtaining the treatment line by subtraction:

	d.f.	s.s.	Mean square
Treatment	7	34 799·0	4971
Residual	48	89 026·1	1855
Total	55	123 825·1	2251

(9.39)

We really require to test the ratio of the residual to the total, but as this is a univariate test we can equally well test $4971/1855 = 2\cdot68$ in the F-distribution with 7 and 48 d.f. This is significant at the 5 percent point. We should infer that the data cannot be wholly explained as an effect on x_2, the grain yield. There appears to be an effect on the straw yield which is obscured if we consider that yield by itself, owing to correlations between the variables.

9.15   This may be a convenient point at which to make some remarks about "degrees of freedom". In univariate theory the term is used in several contexts, beginning with the d.f. of an estimator of variance:

$$\Sigma\,(x - \bar{x})^2/(n - 1). \tag{9.40}$$

This is an unbiased estimate, as compared with the ML estimator in normal samples, which has n as the divisor instead of $n - 1$. The numerator of (9.40) can be transformed by linear functions into the sum of squares of $n - 1$ variables, which again gives some point to the notion of d.f., especially when the splitting of sums of squares in ANOVA is accompanied by the splitting of degrees of freedom. In MANOVA the situation is not so simple. Consider, for example, the dispersion determinant as a generalization of the variance. As we have defined the sample values c_{ij} in the foregoing, they have divisor n, but some writers define them with divisor $n - 1$. This does not mean that the dispersion determinant has d.f. $n - 1$ in any ordinary sense. It may be shown, for example, that the mean value of the dispersion determinant (the c's defined with divisor n) is expressible, so far as normal samples are concerned, in terms of the parent γ by

$$E\{|c|\} = \prod_{j=1}^{p}\left(1 - \frac{j}{n}\right)|\gamma| \qquad (9.41)$$

or, to order n^{-1},

$$= 1 - \frac{p(p+1)}{2n}. \qquad (9.42)$$

It is useful to preserve the classical notation of d.f. if only to remind us of the number of constants fitted. However, in MANOVA our exact tests, as we have seen, are based on an approximation valid for large n, and "corrections" for d.f. are usually of minor importance. Reference may be made to Kendall and Stuart, vol. 3, for the details.

Example 9.5 (data from M.M. Barnard, 1935; Bartlett, 1947)

Miss Barnard had four series of Egyptian skulls, 91 Predynastic, 162 from the sixth to twelfth dynasties, 70 from the twelfth and thirteenth dynasties, and 75 from Ptolemaic dynasties — 398 in all. On each skull four measurements were taken in millimetres:

$$x_1 = \text{maximum breadth}$$
$$x_2 = \text{basi-alveolar length}$$
$$x_3 = \text{nasal height}$$
$$x_4 = \text{basi-bregmatic height}$$

The means of the series are:

	Series I $n_1 = 91$	Series II $n_2 = 162$	Series III $n_3 = 70$	Series IV $n_4 = 75$	
x_1	133·582 418	134·265 432	134·371 429	135·306 667	
x_2	98·307 692	96·462 963	95·857 143	95·040 000	
x_3	50·835 165	51·148 148	50·100 000	52·093 333	
x_4	133·000 000	134·882 716	133·642 857	131·466 667	(9.43)

The sums of squares and products are as follows:

Total (397 d.f.)

	x_1	x_2	x_3	x_4	
x_1	9785·188 442	214·203 518	1217·924 623	2019·831 658	
x_2		9559·459 799	1131·718 593	2381·138 191	
x_3			4088·030 151	1133·467 337	
x_4				9382·253 718	(9.44)

Within series (394 d.f.)

	x_1	x_2	x_3	x_4	
x_1	9662·007 812	445·579 186	1130·619 227	2148·595 591	
x_2		9073·113 822	1239·224 274	2255·824 868	
x_3			3937·618 584	1271·048 148	
x_4				8741·509 700	(9.45)

Between series (3 d.f.)

	x_1	x_2	x_3	x_4	
x_1	123·180 630	−231·375 668	87·305 396	−128·763 933	
x_2		486·345 977	−107·505 681	125·313 323	
x_3			150·411 567	−137·580 811	
x_4				640·734 018	(9.46)

We may first of all consider whether the data are homogeneous, in particular whether there are any significant differences between means of series. The appropriate criterion is that of (9.14), namely the determinant of (9.45) divided by that of (9.44) taken to the $(\frac{1}{2}n)$th power, namely

$$l_{H_2} = \left(\frac{2426·435}{2953·928}\right)^{\frac{1}{2}n} = (0·8214)^{\frac{1}{2}n}.$$

$-2\log_e l_{H_2}$, taking n as the number of degrees of freedom (397), is then 78·1 and the number of d.f. for the approximate χ^2 test is $16 - 4 = 12$. Thus we conclude that the data are not homogeneous, even in the mean.

There are several questions we may wish to ask at this point. For example, from (9.45) and (9.46) it is clear that within series there is a considerable correlation among the variables. Are the differences between the means attributable to influences from all four variables, or, for example, do x_3 and x_4 contribute to the differences only because of their correlation with x_1 and x_2? To answer this we determine the regressions of x_3 and x_4 on x_1 and x_2, extract them from the total variation, and test the residual matrices. Thus we regard x_3 and x_4 as a matrix of dependent variables y and regress them on x_1 and x_2.

In our present case, the dispersion matrix of x_1, x_2 from (9.44) is

$$\frac{1}{394} \begin{bmatrix} 9662·007 812 & 445·579 186 \\ & 9073·113 822 \end{bmatrix} \tag{9.47}$$

the inverse of which is

$$394 \times 10^{-4} \begin{bmatrix} 1·037 331 & −0·050 943 \\ & 1·104 659 \end{bmatrix} \tag{9.48}$$

The variation due to regression of x_3 and x_4 may be expressed, as in **9.16**, as

$$\hat{\beta}v\hat{\beta}' = (uv^{-1})(v)(v^{-1})'u' = uv^{-1}u'.$$

In our case x refers to x_3 and x_4, z to x_1 and x_2, so, using (9.45), we find for this expression

$$10^{-4}\begin{bmatrix} 1130 \cdot 619\,227 & 1239 \cdot 224\,274 \\ 2148 \cdot 595\,591 & 2255 \cdot 824\,868 \end{bmatrix}\begin{bmatrix} 1 \cdot 037\,331 & -0 \cdot 050\,943 \\ -0 \cdot 050\,943 & 1 \cdot 104\,659 \end{bmatrix}$$

$$\times \begin{bmatrix} 1130 \cdot 619\,227 & 2148 \cdot 595\,591 \\ 1239 \cdot 224\,274 & 2255 \cdot 824\,868 \end{bmatrix}$$

$$= \begin{bmatrix} 287 \cdot 966\,771 & 534 \cdot 240\,445 \\ 534 \cdot 240\,445 & 991 \cdot 630\,237 \end{bmatrix}. \tag{9.49}$$

Subtracting this from the matrix of x_3 and x_4, we have as residual

$$\begin{bmatrix} 3649 \cdot 651\,813 & 736 \cdot 807\,703 \\ & 7749 \cdot 879\,463 \end{bmatrix}, \tag{9.50}$$

with $394 - 2 = 392$ d.f.

Similarly, operating on (9.45) for the totals of product sums, we find the residual

$$\begin{bmatrix} 3808 \cdot 643\,292 & 611 \cdot 789\,751 \\ & 8393 \cdot 746\,855 \end{bmatrix}. \tag{9.51}$$

The question is whether the matrices (9.50) and (9.51) are significantly different. We can regard the latter as the residual in the regression of x_3, x_4 on x_1, x_2 plus a vector representing the mean; the former has had the mean abstracted in each class. The criterion is the $(\frac{1}{2}n)$th power of the ratio of the determinants, namely

$$l_{H_2} = \left(\frac{0 \cdot 277\,415}{0 \cdot 315\,944}\right)^{\frac{1}{2}n} = (0 \cdot 8780)^{\frac{1}{2}n}.$$

$-2 \log l$, with $n = 392$, is then 50·98 and the appropriate number of degrees of freedom for $p = 2$, $k = 4$ is $8 - 2 = 6$. Homogeneity is therefore rejected. We conclude that x_3 and x_4 are relevant variables in the sense that the differences between means cannot be ascribed to x_1 and x_2 alone.

A further question considered by Miss Barnard was whether these variables might each have a linear regression on time. To investigate this we require a time variable, and the intervals between the four series were taken proportionately to 2, 1, 2. We may therefore conveniently take the values of t as $-5, -1, 1, 5$. On this basis,

$$S(t - \bar{t})^2 = 4307 \cdot 668\,342$$
$$Sx_1(t - \bar{t}) = 718 \cdot 748\,743$$
$$Sx_2(t - \bar{t}) = -1407 \cdot 271\,358$$
$$Sx_3(t - \bar{t}) = 410 \cdot 100\,502$$
$$Sx_4(t - \bar{t}) = -733 \cdot 442\,212$$

We are now examining the regression of each of the x's on the extraneous variable, time. The sums of squares and products due to regression (1 degree of freedom) are

	x_1	x_2	x_3	x_4
x_1	119·925 610	−234·807 891	68·426 628	−122·377 264
x_2		459·741 214	−133·975 656	239·608 098
x_3			39·042 565	−69·825 482
x_4				124·879 038	(9.52)

Here, for example, the item in row 1 and column 2 is

$$\frac{Sx_1(t - \bar{t})Sx_2(t - \bar{t})}{S(t - \bar{t})^2} = \frac{(718 \cdot 748\,743)(-1407 \cdot 271\,358)}{4307 \cdot 668\,342} = -234 \cdot 807\,891.$$

The residual after removing the regression on time from the original matrix is given by subtracting (9.52) from (9.44), namely

	x_1	x_2	x_3	x_4
x_1	9665·262 832	449·011 409	1149·497 995	2142·208 922
x_2		9099·718 585	1265·694 249	2141·530 093
x_3			4048·987 586	1203·292 819
x_4				9257·364 680	(9.53)

with 396 d.f.

We now test whether this residual is homogeneous against the determinant of (9.45). The ratio of determinants is 0·9031. The criterion is 40 with 8 d.f. (16 means in one case, four means plus four regression coefficients fitted in the second). This is significant. We conclude that if the regression on time is linear, there are differences between the series which are not due to temporal effects.

9.16 We now consider a somewhat more general type of regression, in which the model is

$$\underset{q \times n}{\mathbf{y}} = \underset{q \times p}{\boldsymbol{\beta}} \underset{p \times n}{\mathbf{x}} + \underset{q \times n}{\boldsymbol{\epsilon}}. \qquad (9.54)$$

Thus we have a vector of q y's, p x's and a residual element ϵ, the elements of which may not be independent, i.e. ϵ_j may be dependent on ϵ_k. If this were not so, each component of y could be considered separately.

As in the standard regression situation, we estimate $\boldsymbol{\beta}$ by maximizing the likelihood, which is given by

$$L = \frac{|\alpha|^{\frac{1}{2}p}}{(2\pi)^{\frac{1}{2}np}} \exp\left\{-\tfrac{1}{2} \underset{t=1}{\overset{n}{S}} (y_t - \boldsymbol{\beta} x_t)' \boldsymbol{\alpha} (y_t - \boldsymbol{\beta} x_t)\right\}, \qquad (9.55)$$

where $\boldsymbol{\alpha}$ is inverse to the covariance determinant of $\boldsymbol{\epsilon}$, say σ_{jl}.

For the ML estimator of $\boldsymbol{\beta}$ it is found that

$$\underset{q \times p}{\hat{\boldsymbol{\beta}}} = \underset{q \times p}{\mathbf{u}} \underset{p \times p}{\mathbf{v}^{-1}}, \qquad (9.56)$$

where

$$u_{sk} = y_s x_k' \qquad (9.57)$$

$$v_{sk} = x_s' x_k. \qquad (9.58)$$

It may also be shown that

$$\hat{\sigma}^2 = \frac{1}{n} (\mathbf{y} - \hat{\boldsymbol{\beta}} \mathbf{x})(\mathbf{y} - \hat{\boldsymbol{\beta}} \mathbf{x})'.$$

It is remarkable that we get the same equations (9.56) whether the ϵ's are dependent or not. Thus we can estimate the β's separately for the components of y, treating them as separate problems.

9.17 The internal structure of σ_{ij} appears, however, in the sampling covariances of the β's. In fact, it may be shown that

$$E\,(\hat{\boldsymbol{\beta}}_j - \boldsymbol{\beta}_j)'(\hat{\boldsymbol{\beta}}_l - \boldsymbol{\beta}_l) = \sigma_{jl} \mathbf{v}^{-1}. \qquad (9.59)$$

$\hat{\sigma}^2$ may also be written

$$\hat{\boldsymbol{\sigma}}^2 = \frac{1}{n}(\mathbf{y}\mathbf{y}' - \hat{\boldsymbol{\beta}} \mathbf{v} \hat{\boldsymbol{\beta}}'). \qquad (9.60)$$

9.18 There are a few other standard cases in which the likelihood ratio provides a test of hypotheses. In fact, Hotelling's T^2 can itself be regarded as an l-test with

$$l = \{1 + T^2/(n-1)\}^{\frac{1}{2}n}. \qquad (9.61)$$

Consider the case where we suspect that a set of variables is split into subgroups, each independent of the others but possibly embodying dependences within itself.

If the p variables are divided into subsets of $p_1, p_2, \ldots, p_q$ variables, the covariance matrix may be displayed:

$$\begin{bmatrix} \gamma_{11} & \gamma_{12} & \cdots & \gamma_{1q} \\ \gamma_{21} & \gamma_{22} & \cdots & \gamma_{2q} \\ \cdot & \cdot & \cdots & \cdot \\ \gamma_{q1} & \gamma_{q2} & \cdots & \gamma_{qq} \end{bmatrix} \qquad (9.62)$$

The hypothesis under test is that all the off-diagonal elements in (9.62) vanish. It may be shown that the test l_H has the criterion

$$l_H = \left\{ \frac{|c|}{\prod_{j=1}^{q} |c_{jj}|} \right\}^{\frac{1}{2}n}. \tag{9.63}$$

$-2 \log l$ is distributed as χ^2 with degrees of freedom r, given by

$$r = \tfrac{1}{2}\{p(p+1) - \Sigma\, p_j(p_j + 1)\}. \tag{9.64}$$

In the particular case when all the γ's are unity, so that effectively we are testing for non-correlation, (9.63) becomes the product of the diagonal terms. Consequently (9.63) reduces to the correlation determinant r, tested with $\tfrac{1}{2}p(p-1)$ d.f.

9.19 Likewise, a test that all variables have equal variance and are independent, namely that the parent covariance matrix is $\sigma^2 I$, where I is the identity matrix, has the criterion

$$-\log \left[\frac{|c|}{\left\{ \dfrac{1}{p}\, \text{trace}\, c \right\}^{p}} \right]^{\frac{1}{2}n}, \tag{9.65}$$

distributed approximately as χ^2 with $\tfrac{1}{2}p(p+1) - 1$ d.f. $|c|$ is the product of the eigenvalues, and trace c/p is the average eigenvalue (the trace being their sum), so that on intuitive grounds alone one might expect some such test.

We can apply the result to test whether a sample comes from a parent with covariance matrix γ. In fact, from γ we can find the transformation which reduces it to $\sigma^2 I$. This transformation, applied to the sample, gives us new variables, and their matrix can be tested as at (9.65).

NOTES

(1) The power of the tests developed in this chapter has been investigated by a number of authors, but the whole subject is complicated if we wish to consider power under variation of separate parameters.

(2) A paper by Kullback and Rosenblatt (1957) considers the testing of various hypotheses concerning regression coefficients (e.g. that some of the coefficients are equal).

10

Discrimination

10.1 In Chapter 3 we considered the problem of classifying individuals into separate groups on the basis of their observed characteristics. We now discuss a problem of a fundamentally different kind: given the existence of two populations A and B; given a random sample of individuals which we know for certain come from A, and another random sample which we know for certain come from B; how do we set up a rule to allocate further individuals, of whose origin we are uncertain, to the correct population (assuming that they come from either A or B)? We should like to be able to do this in some optimal sense, e.g. by making as few mistakes as possible, or minimizing the cost of such mistakes as we do make. This is the problem of *discrimination*.

10.2 It is a fair question to ask why a problem exists. If there has been some criterion by which we know for certain the population of origin of the initial samples, why do we not always use it? There are at least four types of case providing an answer to this question.

(1) Lost information. An archaeologist or an anthropologist may require to sex the bones discovered in an ancient burial ground. While the subjects were alive there would have been no problem, and there is plenty of living material on which bone measurements can be made. The information in the archaeological material has gone the way of all flesh.

(2) Unobtainable information. A hospital may have records relating external symptoms to internal conditions requiring operation, or perhaps obtainable only in post mortem. The problem is to diagnose the conditions from external symptoms without such drastic procedures.

(3) Prediction. It may have been found in the past that we can forecast certain phenomena from others which occur previously in time. At the current moment we observe these leading variables, but we wish to forecast the one which depends on them without waiting for it to happen.

(4)   Testing to destruction. When a test (like firing a torpedo) results in the destruction of the individual, we require a method of predicting the performance of other individuals without destroying them all.

10.3   We have stated the problem, and initially shall so consider it, with two restrictions which can later be removed:

(1)   that there are only two populations of provenance — more generally there may be k;

(2)   that we must allocate new individuals to one of the specified populations and are not allowed to say either that it seems unlikely that they emanated from any of the k, or that we reserve judgement because the indications are not decisive.

10.4   Although our measurements are in p dimensions, we can illustrate most of the argument by drawing scatter diagrams in two. Fig. 10.1 shows two

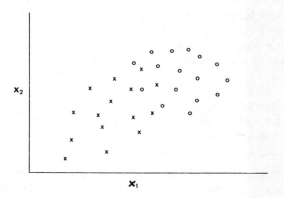

Fig. 10.1

sets of scatter points, represented by crosses and small circles, such as might arise from two parent populations of approximately multivariate normal type. If we are prepared to assume normality we can estimate the parameters of each population and hence estimate the probability (the likelihood function) of any new point such as Q. The criterion we shall adopt is a very simple one. We allocate Q to population A or B according to whether the likelihood of it coming from A is greater or less than the likelihood of it coming from B. If L_A and L_B are the likelihood functions of A and B, there will be a boundary where $L_A = L_B$. Points falling on one side we allocate to A; those on the other to B. We are, as it were, allocating Q to the "nearer" population, except that "nearness" is a measure in probability, not a metrical distance. If both likelihoods are small, we may well doubt whether Q came from either population; and if they are nearly equal, we allocate Q with great diffidence.

10.5   A rather more sophisticated approach is as follows. Let f_1 and f_2 be the (multivariate) frequency functions of A and B, and let us divide up the First Space W into a region R and the remainder $W - R$, such that the probability of misallocation is the same for the two kinds of mistake we may make, namely putting Q in R when it belongs to $W - R$, and vice versa. Then, writing one sign for the multiple integral over p dimensions,

so that
$$\int_R f_2 \, dx = \int_{W-R} f_1 \, dx = 1 - \int_R f_1 \, dx, \tag{10.1}$$

$$\int_R (f_1 + f_2) dx = 1. \tag{10.2}$$

We further wish to minimize the total error, which may be written

$$1 + \int_R (f_2 - f_1) dx = 2 \int_R f_2 \, dx. \tag{10.3}$$

This minimization is for variations of the region R, subject to the constraint (10.2). Taking a Lagrange multiplier λ, we have to find an unconditional minimum of

$$\int_R \{f_2 - \lambda(f_1 + f_2)\} dx. \tag{10.4}$$

This is clearly achieved when R is such that the integrand in (10.4) is never positive, i.e. is such that on the boundary,

$$\lambda f_1 = (1 - \lambda) f_2, \tag{10.5}$$

or the ratio f_1/f_2 is constant. The value of λ is decided from (10.2). If we forego the requirement that errors of the two kinds are equal, it follows from (10.3) that $f_1/f_2 = 1$.

10.6   Now suppose that the populations are multivariate normal with means μ_1 and μ_2 and *identical covariance matrices*. The equality of likelihood is equivalent to the equality of log likelihoods, so for the discriminatory boundary we have

$$-\tfrac{1}{2} \sum_{j,k=1}^{p} \alpha_{jk}(x_j - \mu_{1j})(x_k - \mu_{1k}) = -\tfrac{1}{2} \sum_{j,k=1}^{p} \alpha_{jk}(x_j - \mu_{2j})(x_k - \mu_{2k}). \tag{10.6}$$

The quadratic terms cancel, and this is equivalent (since α_{jk} is symmetric) to

$$\sum \alpha_{jk}\mu_{1j} x_k - \tfrac{1}{2} \sum \alpha_{jk}\mu_{1j}\mu_{1k} = \sum \alpha_{jk}\mu_{2j} x_k - \tfrac{1}{2} \sum \alpha_{jk}\mu_{2j}\mu_{2k}. \tag{10.7}$$

We may also write

$$\sum \alpha_{jk}(\mu_{1j} - \mu_{2j}) x_k = \tfrac{1}{2} \sum \alpha_{jk}(\mu_{1j}\mu_{1k} - \mu_{2j}\mu_{2k}). \tag{10.8}$$

This is the form in which it is usually written for two populations, but generalization to more than two is more symmetrically stated at (10.7). (See below, **10.11**.)

10.7   Equations (10.7) and (10.8) are the parental forms. In practice we have to estimate from the sample, and as usual we find

$$\Sigma\, C_{jk}(\bar{x}_{1j} - \bar{x}_{2j})x_k = \tfrac{1}{2}\, \Sigma\, C_{jk}(\bar{x}_{1j}\bar{x}_{1k} - \bar{x}_{2j}\bar{x}_{2k}), \qquad (10.9)$$

where C_{jk} is inverse to c_{jk}.

10.8   The same result may be reached by a different argument. Suppose we seek a linear function of the x's,

$$X = \sum_{j=1}^{p} l_j x_j, \qquad (10.10)$$

so as to maximize the ratio of between-class to within-class variances (the familiar procedures to "separate" two groups as much as possible). We then have to maximize

$$\frac{\{\Sigma\, l_j(\mu_{1j} - \mu_{2j})\}^2}{\Sigma\, l_j l_k \gamma_{jk}}. \qquad (10.11)$$

A differentiation with respect to l_j gives us

$$\mu_{1j} - \mu_{2j} = \frac{\sum\limits_{j=1}^{p} l_j(\mu_{1j} - \mu_{2j})\,\Sigma\, \gamma_{jk} l_k}{2\,\Sigma\, l_j l_k \gamma_{jk}}, \qquad (10.12)$$

from which we have

$$l_k \propto \sum_{j=1}^{p} \Gamma_{jk}(\mu_{1j} - \mu_{2j}), \qquad (10.13)$$

Γ being inverse to γ. We are led back to the equation of type (10.8).

Example 10.1 (Fisher, 1936)

(Data in Table 3.4.) This famous example has been worked over by many writers. Fisher exemplified the discriminant on *setosa* and *versicolor*, omitting to notice that petal width alone separates the two varieties with virtual certainty.

The three varieties of Iris have the following variables: $x_1 =$ sepal length; $x_2 =$ sepal width; $x_3 =$ petal length; $x_4 =$ petal width. The means were (in cm)

Variate	Versicolor	Virginica	Difference, Ve − Vi
x_1	5·9360	6·5880	−0·6520
x_2	2·7700	2·9740	−0·2040
x_3	4·2600	5·5520	−1·2920
x_4	1·3260	2·0260	−0·7000 (10.14)

The pooled sums of squares and products about the respective means (in cm^2) were 100 times

	x_1	x_2	x_3	x_4
x_1	0·3287	0·0877	0·2382	0·0514
x_2		0·0992	0·0755	0·0435
x_3			0·2574	0·0597
x_4				0·0561

(10.15)

The inverse matrix is, in cm^{-2},

	x_1	x_2	x_3	x_4
x_1	9·8851	−3·2760	−8·8663	2·9280
x_2		17·4597	0·4770	−11·0494
x_3			13·4812	−6·6025
x_4				30·7346

(10.16)

Questions of degrees of freedom sometimes arise in this class of work, but since the object of the discrimination function is to separate, it can absorb an arbitrary constant in the coefficients. We shall take the total sample number 100 as the number of degrees of freedom in (10.15).

Using (10.16), we then find for the coefficients

$$l_1 = (9·8851)(-0·652) - 3·2760(-0·204) - 8·8663(-1·292)$$
$$+ 2·9280(-0·700) = 3·6289$$
$$l_2 = 5·6925, \quad l_3 = -7·1124, \quad l_4 = -12·6388.$$

(10.17)

Our discriminant is then

$$X = 3·6289x_1 + 5·6925x_2 - 7·1124x_3 - 12·6388x_4.$$ (10.18)

The constant on the right in (10.9) is one-half of the difference between X when we substitute the means of $\bar{x}_i$ for x_i from *versicolor* and the similar quantity for *virginica*, namely

$$\tfrac{1}{2}\{-9·7485 - (-24·2576)\} = 7·2546.$$

We allot to *versicolor* or *virginica* for a new individual according as X falls below or exceeds this figure.

10.9   We may calculate approximately the probability of misclassification. We have

$$\text{var } X = \Sigma \, l_j l_k \gamma_{jk}$$
$$= \Sigma \, l_j \gamma_{jk} \Gamma_{km}(\mu_{1m} - \mu_{2m})$$
$$= \Sigma \, l_j(\mu_{1j} - \mu_{2j})$$

which is estimated by

$$\text{est. var } X = \bar{X}_1 - \bar{X}_2.$$ (10.19)

Example 10.2

In the data of Example 10.1,

$$\bar{X}_1 - \bar{X}_2 = -9 \cdot 7485 + 24 \cdot 2576 = 14 \cdot 5084.$$

The square root is $3 \cdot 809$. The probability of misclassification (assuming normality, so that the linear function of (10.18) is normal) is then that of a deviation of $7 \cdot 2546$ in a normal distribution with zero mean and standard deviation $3 \cdot 809$, i.e. a deviation of $1 \cdot 90$, yielding a probability of about $0 \cdot 05$, or a 5 percent error.

With the aid of a computer we can estimate the probability of misclassification empirically. In fact, if we apply the discriminant to each of the observations on which it was based, we find that in *versicolor* observations 21 and 27 would be misclassified, and in *virginica* observation 34 would also be misclassified. Altogether the misclassification error is 3 percent. We should in any case expect this to be somewhat lower than the true probability, because we have fitted the discriminant to these actual data.

10.10   The assumption that the parent populations are not only multivariate normal but have a common dispersion matrix is obviously a very drastic one, which we should be lucky to encounter in practice. Consider first of all the abandonment of the requirement as to a common dispersion matrix. The equality of log likelihood is equivalent to

$$\tfrac{1}{2} \log |\alpha_1| - \tfrac{1}{2} \Sigma \, \alpha_{1jk}(x_{1j} - \mu_{1j})(x_{1k} - \mu_{1k})$$
$$= \tfrac{1}{2} \log |\alpha_2| - \tfrac{1}{2} \Sigma \, \alpha_{2jk}(x_{1j} - \mu_{2j})(x_{2k} - \mu_{2k}). \quad (10.20)$$

The discriminating boundary is then quadratic and is an awkward construct to handle in more than two dimensions. For this reason efforts have been made to avoid quadratic boundaries

(1) by reducing the discriminator to two dimensions;
(2) by seeking distribution-free methods;
(3) by empirical studies of the patterns of behaviour thrown up by the computer on a visual display unit.

Example 10.3

Biometricians have often proposed to discriminate between individuals on the basis of "size" and "shape". Consider the case where measurements are made on an organism and the correlations between them are positive and all equal, say, to r. The correlation matrix is then the one that we considered in Example 1.4 and has eigenvalues $\lambda_1 = 1 + (p-1)r$, $\lambda_2 = \ldots = \lambda_p = 1 - r$. From the component analysis approach this means that there is one principal direction and that the others are isotropic. The component corresponding to λ_1 is

$$\xi_1 = \frac{1}{\sqrt{p}} \Sigma x_i. \tag{10.21}$$

We take a "size" component proportional to this and write

$$Q = \Sigma x_i = \xi_1\sqrt{p}, \tag{10.22}$$

so that

$$\text{var } Q = p\lambda_1 = p\{1 + (p-1)r\}. \tag{10.23}$$

Among the remaining variation no particular direction is suggested as suitable. Let us take a set of weights w_i with non-zero mean w and define a shape component by

$$P = \sum_{j=1}^{p} \frac{w_j - w}{w} x_j. \tag{10.24}$$

We have at once

$$\text{var } P = (1-r) \Sigma \left(\frac{w_j - w}{w}\right)^2, \tag{10.25}$$

where the x's are taken to have standard measure. Also

$$\begin{aligned}
\text{cov } (Q, P) &= \text{cov } \left\{\Sigma \frac{w_j - w}{w} x_j, \; \Sigma x_j\right\} \\
&= \sum_j \frac{w_j - w}{w} \text{var } x_j + \sum_{i \neq j} \frac{w_j - w}{w} \text{cov } (x_i, x_j) \\
&= \{1 + (p-1)r\} \Sigma \frac{w_j - w}{w} \\
&= 0. \tag{10.26}
\end{aligned}$$

The shape component is then uncorrelated with the size component, and this will remain approximately true if the correlations are nearly equal but not exactly so.

When we are interested in discrimination we may choose the weights so as best to discriminate between two populations and hence arrive at an *ad hoc* measure of "shape". We shall take

$$w_j = \bar{x}_{1j} - \bar{x}_{2j} \tag{10.27}$$

and shall look for a linear function of size and shape which is the best discriminator:

$$X = aQ + P. \tag{10.28}$$

This will be given by determining a so as to maximize

$$\frac{(X_1 - X_2)^2}{\text{var } X}$$

If $D_p = P_1 - P_2$ and $D_Q = Q_1 - Q_2$, the suffixes as usual referring to the two populations, this requires the maximization of

$$\frac{(aD_Q + D_P)^2}{a^2 \operatorname{var} Q + 2a \operatorname{cov}(Q,P) + \operatorname{var} P},$$

leading to

$$a = \frac{D_Q \operatorname{var} P - D_P \operatorname{cov}(Q,P)}{D_P \operatorname{var} Q - D_Q \operatorname{cov}(Q,P)}. \tag{10.29}$$

But Q and P are uncorrelated. We have also, using sample means as estimates of parent means,

$$D_P = \sum_{j=1}^{p} \frac{\bar{x}_{1j} - \bar{x}_{2j} - (\bar{x}_1 - \bar{x}_2)}{\bar{x}_1 - \bar{x}_2}(x_{1j} - x_{2j})$$

$$= \sum \left\{ \frac{(\bar{x}_{1j} - \bar{x}_{2j})^2}{\bar{x}_1 - \bar{x}_2} - (\bar{x}_1 - \bar{x}_2)^2 \right\}.$$

$$\operatorname{var} P = (1-r) \sum \left(\frac{x_{1j} - x_{2j} - (x_1 - x_2)}{\bar{x}_1 - \bar{x}_2} \right)^2$$

$$= (1-r) \sum \left\{ \frac{(\bar{x}_{1j} - \bar{x}_{2j})^2}{(\bar{x}_1 - \bar{x}_2)^2} - p(\bar{x}_1 - \bar{x}_2) \right\}$$

$$D_Q = p(\bar{x}_1 - \bar{x}_2)$$

$$\operatorname{var} Q = p\{1 + (p-1)r\}.$$

On substitution for a in (10.29) we then find

$$X = \frac{1-r}{1+(p-1)r}Q + P. \tag{10.30}$$

This, in the sense we have defined, is the "best" linear function of size and shape.

Example 10.4 (C.A.B. Smith, 1947)

The method of the foregoing example is useful in reducing the discriminant to a function of two variables only, and if we work on size and shape variables P and Q we can handle quadratic discriminators without too much mathematical complexity.

In the present example we have, for x and y, variables which are already constructed as size and shape variables from the original data. Their correlations are not significant.

A group of 25 normal and 25 psychotic individuals were given certain tests, and for each individual a size and shape variable x and y were determined. The results were:

	25 Normals			25 Psychotics	
x	y	z	x	y	z
22	6	62	24	38	8
20	14	36	19	36	-13
23	9	61	11	43	-67
23	1	77	6	60	-126
17	8	33	9	32	-55
24	9	66	10	17	-20
23	13	53	3	17	-55
18	18	18	15	56	-73
22	16	42	14	43	-52
19	18	23	20	8	48
20	17	30	8	46	-88
20	31	2	20	62	-60
21	9	51	14	36	-38
13	13	3	3	12	-45
20	14	36	10	51	-88
19	15	29	22	22	30
20	11	42	11	30	-41
18	17	20	6	30	-66
20	7	50	20	61	-58
23	6	67	20	43	-22
23	23	33	15	48	-57
25	9	71	5	53	-117
23	5	69	10	43	-72
21	12	45	13	19	-9
23	7	65	12	4	16
Totals 520	308	1084	320	910	-1120
Mean 20·80	12·32	43·4	12·80	36·40	$-44·8$ (10.31)

Here z is a quantity $5x - 2y - 36$ to be explained later. We have the following quantities:

	Normal	Psychotics
Mean of x	20·80	12·80
Mean of y	12·32	36·40
var x	6·92	36·75
var y	40·89	287·92
cov (x, y)	$-5·27$	13·92
d.f.	24	24
s.d. of x	2·63	6·06
s.d. of y	6·39	16·97
Correlation	$-0·31$	0·14

(10.32)

Let us consider first of all a linear discriminant function of x and y. Pooling the dispersions, we get for the supposed dispersion matrix

$$
\begin{array}{cc}
 & \begin{array}{cc} x & y \end{array} \\
\begin{array}{c} x \\ y \end{array} &
\begin{bmatrix} 21\cdot83 & 4\cdot33 \\ & 164\cdot40 \end{bmatrix}
\end{array}
\qquad (10.33)
$$

and for the difference of means (Normal $-$ Psychotics)

$$
\begin{array}{cc}
x & 8\cdot00 \\
y & -24\cdot08
\end{array}
$$

The inverse of (10.33) is proportional to

$$
\begin{bmatrix} 164\cdot40 & -4\cdot33 \\ -4\cdot33 & 21\cdot83 \end{bmatrix}
$$

This discriminant is then

$$
(164\cdot40 \times 8\cdot00 + 4\cdot33 \times 24\cdot08)\ \{x - \tfrac{1}{2}(20\cdot80 + 12\cdot80)\}
$$
$$
+ (-4\cdot33 \times 8\cdot00 + 21\cdot83 \times -24\cdot08)\ \{y - \tfrac{1}{2}(12\cdot32 + 36\cdot40)\}
$$
$$
= 1419x - 560y - 10\,198.
$$

On division by 280 this becomes, nearly enough,

$$
z = 5x - 2y - 36. \qquad (10.34)
$$

The values of this function are shown in (10.31). It is seen that z is positive for all normals (no errors) and negative for all psychotics except four (16 percent error). The errors of classification, as estimated from the data themselves, are accordingly not symmetrical and amount to 8 percent over all. This is better than we should do by using x or y alone; for instance, if we take $x \geqslant 17$ to be normal and $x \leqslant 16$ to be psychotic, there would be one error in classifying the normals and six for the psychotics.

We may remark, however, that the variances and covariances of x and y are very different for the two types; and we doubt whether it is legitimate to suppose that they have a common dispersion matrix. If we assume different dispersion matrices $\boldsymbol{\alpha}$ and $\boldsymbol{\beta}$, say, the logarithm of the likelihood ratio becomes proportional to

$$
\Sigma\, \alpha^{ij}(x_i - \mu_{1i})(x_j - \mu_{1j}) - \Sigma\, \beta^{ij}(x_i - \mu_{2i})(x_j - \mu_{2j}), \qquad (10.35)
$$

which is a quadratic function. We now use the fact that size and shape have a correlation which can be put equal to zero. The expression (10.35) then reduces to a form of type

$$
(\zeta - k_1)^2 + (\eta - k_2)^2 + k_3, \qquad (10.36)
$$

where

$$
\zeta = x\sqrt{(\alpha^{11} - \beta^{11})}, \text{ etc.}
$$

In our present case we have for the estimates of **α**, **β**,

$$a^{11} = \frac{1}{6\cdot92} = 0\cdot1445$$

$$a^{22} = \quad\quad 0\cdot0244$$

$$b^{11} = \quad\quad 0\cdot0272$$

$$b^{22} = \quad\quad 0\cdot0035$$

and the discriminant function becomes

$$-0\cdot1173\,(x - 22\cdot65)^2 - 0\cdot0209\,(y - 8\cdot29)^2 + 8\cdot16. \qquad (10.37)$$

On multiplication by -2 this becomes, nearly enough,

$$X = \left(\frac{x - 23}{2}\right)^2 + \left(\frac{y - 8}{5}\right)^2 - 16. \qquad (10.38)$$

The values given to the 25 normals by this function are $-16, -12, -16,$ $-14, -7, -16, -15, -6, -13, -8, -10, 7, -15, 10, -12, -10, -13,$ $-7, -14, -16, -7, -15, -16, -14, -16$. (Two positive, error = 8 percent.)

The values given to the psychotics are $20, 19, 69, 164, 56, 29, 87, 92,$ $53, -14, 95, 103, 36, 85, 100, -8, 39, 76, 99, 41, 64, 146, 75, 14, 5.$ (Two negative, error = 8 percent.)

It is instructive to plot the data (as Smith does) on a diagram and to examine how the points lie in relation to the discriminatory lines.

An example of a quadratic discriminant of zygosity from fingerprints is given by Slater *et al.* (1964).

10.11   The extension of the foregoing to linear discrimination among $k > 2$ populations is made without difficulty. For example, with three populations having common covariance matrix γ we compare the functions, as in (10.7),

$$X_l = \Sigma\,\alpha_{jk}\mu_{lj}x_k - \tfrac{1}{2}\Sigma\,\alpha_{jk}\mu_{lj}\,\mu_{lk}, \quad l = 1, 2, 3, \qquad (10.39)$$

and allot the individual to whichever population gives the greatest value of X. As usual, of course, the parameters are to be estimated from the sample. Extensions to nonlinearity are also possible but rather complicated and should be avoided whenever possible.

For more than three discriminators the division of the sample space may be complicated. Consider, for example, five points in a plane, $A, B, C, D, E,$ and the problem of dividing the plane into five areas a, b, c, d, e so that every point in a is closer to A than to the other four points; and so for $b,$ c, d, e. Such a division might be as in Fig. 10.2.

In more than two dimensions the delimitation of spaces makes a considerable call on one's powers of geometrical imagination. But it is not necessary

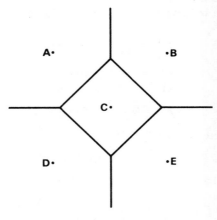

Fig. 10.2

to form a picture. The calculation of the functions such as (10.39) is sufficient.

Example 10.5 (Rao and Slater, 1949)

A number of persons falling into certain neurotic groups obtained the following mean scores in three tests, x_1, x_2, x_3.

Group	Sample size	Mean score 1	2	3
Anxiety state	114	2·9298	1·1667	0·7281
Hysteria	33	3·0303	1·2424	0·5455
Psychopathy	32	3·8125	1·8438	0·8125
Obsession	17	4·7059	1·5882	1·1176
Personality change	5	1·4000	0·2000	0·0000
Normal	55	0·6000	0·1455	0·2182
	256			(10.40)

The dispersion matrix within groups (250 d.f.) was

	1	2	3
1	2·300 851	0·251 578	0·474 169
2		0·607 466	0·035 774
3			0·595 094

(10.41)

Its inverse is

	1	2	3
1	0·543 234	−0·200 195	−0·420 813
2		1·725 807	0·055 767
3			2·012 357

(10.42)

The functions of type (10.39) then are as follows:

Group	Coefficients			Constant
	x_1	x_2	x_3	
Normal	0·2050	0·1431	0·1947	−0·0931
Personality change	0·7204	0·0649	−0·5780	−0·5107
Anxiety state	1·0515	1·4676	0·2974	−2·5047
Hysteria	1·1678	1·5679	−0·1081	−2·7139
Psychopathy	1·3599	2·4641	0·1336	−4·9182
Obsession	1·7680	1·8611	0·3573	−5·8375

(10.43)

Here, for example, the coefficient of x_1 for the normal state is

$$(0·543\,234)\,(0·6000) - (0·200\,195)\,(0·1455) + (-0·420\,813)\,(0·2182)$$
$$= 0·2050.$$

Suppose, for example, we had a subject with scores 1, 1, 0. The values of the functions, in the order of (10.43), are 0·2550, 0·2746, 0·0144, 0·0218, −1·0942, −2·2084. We assign the member to the second group, personality change. In practice, of course, we should do so very tentatively. The normal group is very close, and there are only five members in the personality-change group on which the sample discriminators are based.

10.12   We can now pick up a point which was left over in Chapter 7 concerning the use of dummy regressand variables. For two populations A and B let us score α for A and β for B, and consider the "regression" equation

$$y = \sum_{j=1}^{p} l_j x_j. \qquad (10.44)$$

If there are n_1 members from A and n_2 from B, it is convenient to take the mean score as zero, so that

$$n_1\alpha + n_2\beta = 0. \qquad (10.45)$$

Equivalently we may score kn_2 for A and $-kn_1$ for B, k being an arbitrary constant. The least-squares equations for estimating the l's then become

$$\frac{1}{n}(\Sigma_A\, kn_2 x_i - \Sigma_B\, kn_1 x_i) = \sum_{j=1}^{p} c_{ij} l_j, \qquad (10.46)$$

where, as usual, c_{ij} is the covariance of x_i and x_j and $n = n_1 + n_2$. This gives us, in an obvious notation,

$$l_i = \frac{kn_1 n_2}{n} \sum_j C_{ij}(\bar{x}_{jA} - \bar{x}_{jB}), \qquad (10.47)$$

leading back to the discriminant function.

For more than two groups the argument breaks down. If we have three populations, scoring for them α, β, γ, subject to

$$n_1\alpha + n_2\beta + n_3\gamma = 0, \tag{10.48}$$

the estimating equations reduce to

$$\sum_{j=1}^{p} l_j c_{ij} = \frac{1}{n}\{\alpha n_1(\bar{x}_{iA} - \bar{x}_{iB}) + \gamma n_3(\bar{x}_{iB} - \bar{x}_{iC})\}. \tag{10.49}$$

To make this determinate we have to allot relative values to α and γ. But this is equivalent to assigning an arbitrary degree of separation of the variables.

10.13 Up to this point we have assumed that errors of misclassification are equally important; for example, with two populations, it is as important not to allot to A a member which truly belongs to B, or vice versa. In practical cases this symmetry may not exist — it may be less serious to diagnose the existence of a disease where it does not exist (because treatment will in general reveal the truth) than to conclude that it is not present when, in reality, it is (because the victim may unknowingly develop the condition to serious or even fatal termination).

We can allow for this effect, provided that we can put numerical values on the costs of misclassification. Consider two populations and let the costs of misclassification be c_1 and c_2. Then in place of (10.3) we wish to minimize

$$1 + \int_l (c_2 f_2 - c_1 f_1) dx, \tag{10.50}$$

and it may be shown that the discriminating boundary is not now $f_1 - f_2$ but $c_1 f_1 = c_2 f_2$.

The same argument applies to more than two populations, except (a case rarely encountered) where misclassification costs more if a member is misassigned to one population than to another.

10.14 In the same manner we can make allowance for the case where the probability of a new member arising for assignment may vary according to the population of origin. The situation here depends on how the samples are chosen. Suppose that, on the basis of certain symptoms, we have constructed a discriminator for a certain rare disease. If now a new individual appears *at random*, there is a substantial prior probability that he is not suffering from it, simply because it *is* a rare condition. This may affect the probabilities of misclassification (though, perhaps, it ought not to affect the care we take in diagnosis). Again, considering two populations, let π_1 and π_2 be the probabilities that the new member comes from one population or the other. If c_1 and c_2, as before, are the costs of misclassification, we now have to minimize

$$c_2 \int_R \pi_2 f_2 \, dx + c_1 \int_{W-R} \pi_1 f_1 \, dx = c_1 \pi_1 + \int_R (c_2 \pi_2 f_2 - c_1 \pi_1 f_1) dx$$

$$(10.51)$$

and the dividing line is now determined by $c_1 \pi_1 f_1 = c_2 \pi_2 f_2$. The extension to more than two populations is immediate.

10.15 From the geometrical viewpoint, the discriminating functions represent hypersurfaces in p dimensions. Those of **10.6** are, of course, planes. They are not orthogonal to the lines joining the means of distributions. When we have more than two populations the means will not, in general, be collinear. We might, however, find the line of closest fit to the k means, and use variation in the direction of that line as a discriminator. And in fact, if we have k populations, we may seek a function X given by

$$X = \sum_{j=1}^{p} l_j x_j,$$

such that the ratio of variances between and within classes is maximized. It comes to the same thing to maximize the ratio between classes to total variance, as in **10.8**. If **A** represents the dispersion matrix between classes and **B** the total, this is equivalent to maximizing

$$\lambda = \frac{\sum A_{jk} l_j l_k}{\sum B_{jk} l_j l_k}, \qquad (10.52)$$

which leads to

$$\sum_k (A_{jk} - \lambda B_{jk}) l_k = 0. \qquad (10.53)$$

Thus the largest eigenvalue of $|A - \lambda B| = 0$ provides our discriminator. For details reference may be made to Bartlett (1951a), Williams (1952), and Blackith (1960). An example is given by Oldham and Rossiter (1965) on pneumoconiosis. It appears to me that, in general, the use of one function for discrimination among several populations may be rather Procrustean, unless they are so separate that almost any method will yield reasonable results.

Bias in the estimation of misclassification errors

10.16 The simplest way of estimating errors of misclassification is to apply the observed discriminator to each member of the sample on which it is based, and to observe the errors in that sample. If we were certain about the parent normality, equality of dispersions, and the accuracy of estimators of means and dispersions used in constructing the actual discriminator, we could estimate the errors theoretically as in Example 10.2. But if we are uncertain about the extent to which our discriminator is sensitive to these assumptions, it is better to ascertain the errors in applying it to the observed sample. In fact, this is a procedure we should probably wish to follow in any case, as a precautionary check. It may, however, involve a small bias.

10.17 There are, in practice, two sources of error in this kind of empirical determination of the misclassification error. First of all, we do not know the parent parameters, and our discriminant is based on estimates from the sample. On the average we might expect that for this reason our estimate of error will be greater than the true value. On the other hand, our empirical estimate is derived from data to which it has been fitted. Consequently the empirical estimate will, on the average, be less than it would have been had the discriminator been applied to a new sample: but this itself, as we have seen, would be greater than the true value. How far the two effects cancel out it is hard to determine. In practice the resultant bias does not seem to be large. For a discussion of the subject see Hills (1966).

Redundant variables: standard errors

10.18 It is natural to inquire whether all the variables x which appear in our discriminator are necessary. In Example 10.2 it looks as if they are; in the data discriminating between *setosa* and *versicolor* they are certainly not, petal width alone acting as a practically perfect discriminator. One expects that discarding variables will weaken the discriminatory power, but the loss may be negligible. Looked at from the geometrical viewpoint, if our constellation of points in a p-dimensional space is satisfactorily divided into two by the discriminating hyperplane, the same may be true if we project on to one of the co-ordinate hyperplanes, in which case the variable orthogonal to that plane is redundant.

There are several ways of approaching this problem. It would save a good deal of trouble if we could discard unrewarding variables at the outset without bringing them into the analysis. This, however, is a hazardous operation and is not to be recommended. Elashoff *et al.* (1967) have shown that the simple rules emerging from some studies by Cochran (1964) fail to hold, even when choosing the best two of p dichotomous variables. A more direct approach would be to estimate the misallocation errors by omitting certain variables, but this is apt to be tedious if the number of variables is large.

10.19 Kendall and Stuart (volume 3, chapter on Discrimination) give formulae for the standard errors of coefficients in a linear discriminant. For two populations

$$\text{var } l_j = \left(\frac{1}{n_1} + \frac{1}{n_2}\right) C_{jj} + \frac{l_j^2}{n_1 + n_2} + \frac{1}{n_1 + n_2}(\bar{X}_1 - \bar{X}_2)C_{jj}. \quad (10.54)$$

The use of standard errors may be suggestive of "weak" contributions, but, as we found in the parallel problem in regression in Chapter 7, the discarding of variables when there is near-collinearity present is a rather hazardous operation. The subject would repay further study. One useful possibility would be to transform to principal components which are uncorrelated, and to consider the coefficients in the discriminator when so expressed.

Qualitative data

10.20   Our discussion so far has been in terms of measured variables x. In practice we frequently have to deal with situations where some or all of the variables are qualitative. Let us consider the case where they are all qualitative. Suppose there are p of them, and that the jth variable is divided into s_{jt} categories. In the given sample there will be, say, n_{1jk}, n_{2jk} members in the kth category of the jth variate. If n_1, n_2 are the total sample members, we allot a new member in that sub-class to population 1 or population 2 according as

$$\frac{n_{1jk}}{n_1} \gtreqless \frac{n_{2jk}}{n_2}. \tag{10.55}$$

In short, only the proportions in the class (j, k) are relevant. All the other class frequencies tell us nothing about membership of that class.

10.21   This seems a crude method of procedure, but it is in line with the criterion we adopted for measured variables; for equation (10.55) merely says that we allocate a new member to the class for which it has the greater probability of occurrence.

If the categories s_{jk} can be ordered in k, that is to say if they follow a natural sequence $s_{j1}, s_{j2}, \ldots, s_{jt}$ (as for example in an ordered categorization), it might be possible to utilize information from cells outside the (j, k)th. So far as we know, this has not been attempted.

If we are prepared to prescribe misclassification costs, a somewhat more sophisticated discriminator can be set up on the criterion that a number is to be allocated so as to minimize the cost of misclassification over the whole table. Cochran and Hopkins (1961) examine the procedure; see also Linhart (1959). Hills (1967) discusses a number of "nearest-neighbour" procedures and step-by-step methods.

10.22   Perhaps the most troublesome case is the one in which some variables are measured and some are qualitative. There appears as yet to be no satisfactory theory to deal with this situation. A rather heuristic approach is to construct a score from the qualitative variables (e.g. by representing a dichotomy by 0, 1, a tritomy by $-1, 0, 1$, etc. and averaging over variables) and to use that score as a measured variable in conjunction with the other measured variables. Alternatively, a separate discriminator can be constructed from the measured variables for each cell of the qualitative classification − a tedious procedure and one which is apt to reduce the sample numbers for

each discriminator to a very low point of reliability. The subject would repay
further study.

Reserved judgement

10.23 In many — perhaps most — problems in discrimination it is wise to
allow for reserved judgement in borderline cases, and not to insist on an allocation to one of two classes. This means, in geometrical terms, that we wish
to divide the sample space into three regions R_1, R_2 and D_{12}. If a member
falls into R_1 we allocate it to population 1; if it falls into R_2, to population
2. If it falls into D_{12} we admit that the data are insufficient to make a satisfactory judgement. This region, in general, will contain members of both
populations fairly intimately mixed up together, and in practice we should
probably seek some other criterion to disentangle them.

It is not difficult to use the linear discriminator to set up the region D_{12}.
We merely have to decide on what misclassification probabilities are tolerable,
define R_1 and R_2 in terms of them, and assign D_{12} to the remainder of the
sample space.

10.24 With more than two populations the number of regions becomes
more numerous. With three, for example, we may define regions of doubt
D_{12}, D_{23}, D_{31} in terms of the three discriminants, but these will intersect.
Thus we may have a region D_{123} wherein we cannot allocate to any population; a region $D_{12.3}$ where we can reject R_3 but cannot allocate as between
R_1 and R_2; and so on. No particular difficulty arises, at least with linear
discriminators, which divide the sample space into regions with flat boundaries. Problems of interpretation could arise with quadratic or cubic forms.

10.25 As the classical theory of discrimination has developed, at least in
the hands of statisticians, it has been directed towards finding a discriminant
function. However, it is not necessary to condense the operation into a single
discriminatory quantity. Any method which satisfactorily separates the classes
will serve. We proceed to discuss a method which, although perhaps not optimal, is distribution-free and involves only elementary arithmetic.

Let us revert to the representation of members as points in a p-dimensional
space whose co-ordinates are the values of the variables $x_1, x_2, \ldots, x_p$. Con-
fining ourselves for the present to two populations, we may think of one
population (say A) as represented by crosses and the other (say B) by circles.
In two dimensions the picture might look like Fig. 10.3. The crosses have a
convex hull which we have drawn in; likewise for the circles. In general these
two will have a common domain.

Let us consider the following rule of discrimination:

(a) if a point falls in the A-hull but not in the B-hull we assign it to A;
(b) if the point falls in the B-hull but not in the A-hull we assign it to B;

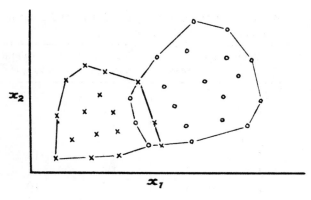

Fig. 10.3 (see text)

(c) if the point falls into both hulls we will not assign it to either.

The proposal is plausible but we shall not follow it up for three reasons:

(i) the determination of the convex hulls is a problem in linear programming which is soluble but takes us outside our present scope;
(ii) the method gives no guide to the treatment of new points which fall outside both hulls;
(iii) the method is not truly distribution-free, because nonlinear variate transformations do not preserve the planarity of the hull boundaries.

A count of points in the two hulls and their common part is nevertheless useful as giving us a measure of the degree of entanglement of the two populations — a measure, so to speak, of the magnitude of the discrimination problem.

10.26 The method we propose may be illustrated on the discrimination of *versicolor* against *virginica*. A casual inspection of the data shows what can be confirmed by tabulation, i.e. that the two differ more on petal length PL and petal width PW than on sepal length or width. We form a frequency distribution for PL and PW as in Table 10.1.

We observe that on PL the two distributions overlap in the range 4·5–5·1. Outside this range there are 29 cases of *versicolor* and 34 cases of *virginica*. On PW there is overlap in the range 1·4–1·8, 28 cases of *versicolor* and 34 of *virginica* lying outside it. The total of cases lying outside the common range being 63 for PL and 62 for PW, we shall take as our first discriminating variable PL.

We then lay down the following rule of discrimination (10.56).

Table 10.1 *Frequency distributions of petal length and petal width for Iris versicolor and Iris virginica*

Variate values	Petal length Vers.	Petal length Virg.	Variate values	Petal width Vers.	Petal width Virg.
⩽ 4·3	25				
4·4	4		1·0	7	
4·5	7	1	1·1	3	
4·6	3	—	1·2	5	
4·7	5	—	1·3	13	
4·8	2	2	1·4	7	1
4·9	2	3	1·5	10	2
5·0	1	3	1·6	3	1
5·1	1	7	1·7	1	1
5·2		2	1·8	1	11
5·3		2	1·9		5
5·4		2	2·0		6
5·5		3	2·1		6
5·6		6	2·2		3
5·7		3	2·3		8
5·8		3	2·4		3
⩾ 5·9		13	2·5		3
	50	50		50	50

Table 10.2 *Frequency distribution of 37 cases not distinguished by PL*

Variate values	Petal width Vers.	Petal width Virg.
1·2	1	
1·3	2	
1·4	4	
1·5	9	2
1·6	3	—
1·7	1	1
1·8	1	5
1·9		3
2·0		3
2·1		—
2·2		—
2·3		1
2·4		1
	21	16

$$PL \leqslant 4.4 \qquad \text{allot to } \textit{versicolor}$$
$$PL \geqslant 5.2 \qquad \text{allot to } \textit{virginica}$$
$$4.5 \leqslant PL \leqslant 5.1 \quad \text{refer to next variable.} \qquad (10.56)$$

There are 37 cases for which PL lies in the common range 4·5–5·1. We take these cases out of Table 10.1 and construct a distribution for them in respect of PW, as in Table 10.2.

Proceeding as before, we see that there is a common range for PW of 1·5–1·8. We therefore add to the rule (10.56)

$$4.5 \leqslant PL \leqslant 5.1$$

$$PW \leqslant 1.4 \qquad \text{allot to } \textit{versicolor}$$
$$PW \geqslant 1.9 \qquad \text{allot to } \textit{virginica}$$
$$1.5 \leqslant PW \leqslant 1.8 \quad \text{proceed to next variable} \qquad (10.57)$$

This leaves 22 cases undecided. PL has discriminated 63 cases and PW a further 15. We now refer to the 22 undecided cases on sepal length SL and sepal width SW (Table 10.3).

Table 10.3 *Frequency distributions of 22 cases not distinguished by PL and PW*

Variate values	Sepal length		Variate values	Sepal width	
	Vers.	*Virg.*		*Vers.*	*Virg.*
4·9		1	2·2	1	1
—	—	—	2·3	—	—
5·4	1	—	2·4	—	—
5·5	—	—	2·5	1	1
5·6	1	—	2·6	—	—
5·7	—	—	2·7	1	1
5·8	—	—	2·8	1	2
5·9	1	1	2·9	1	—
6·0	3	2	3·0	3	3
6·1	—	1	3·1	2	
6·2	1	1	3·2	2	
6·3	2	2	3·3	1	
6·4	1	—	3·4	1	
6·5	1	—			
6·6	—	—			
6·7	2	—			
6·8	—	—			
6·9	1	—			
	14	8		14	8

For SL there are only five cases out of 22 lying outside the common range. For SW there are six. We take SW as out next discriminator and add to (10.57)

$4\cdot5 \leqslant PL \leqslant 5\cdot1$

$1\cdot5 \leqslant PW \leqslant 1\cdot8$

SW $\geqslant 3{:}1$ allot to *versicolor*

SW $< 3\cdot1$ proceed to next variable   (10.58)

Our third variable discriminates a further six, making 84 altogether and leaving 16 undecided. For these 16 the distribution on SL is given in Table 10.4.

Table 10.4   *Distribution of 16 cases not distinguished by PL, PW, SW*

Variate values	Sepal length	
	Vers.	Virg.
4·9		1
—	—	—
5·4	1	—
5·5	—	—
5·6	1	—
5·7	—	—
5·8	—	—
5·9	—	1
6·0	2	2
6·1	—	1
6·2	1	1
6·3	1	2
6·4	—	—
6·5	1	—
6·6	—	—
6·7	1	—
	8	8

For what it is worth we may now add to (10.58)

$4\cdot5 \leqslant PL \leqslant 5\cdot1$

$1\cdot5 \leqslant PW \leqslant 1\cdot8$

SW $< 3\cdot1$

SL $\geqslant 6\cdot4$ allot to *versicolor*

SL $\leqslant 5\cdot3$ allot to *virginica*

$5\cdot4 \leqslant SL \leqslant 6\cdot3$ undecided   (10.59)

This leaves us with 87 cases decided and 13 undecided. No further discrimination is possible.

10.27   The general method will now be clear. It is completely distribution-free, depending only on the rank order of the variate values. It brings up one by one the variables which are *prima facie* most important in the discrimination. It involves no arithmetic other than counting.

On the other hand, the discrimination which results is not necessarily optimal. Looked at from the geometrical viewpoint, instead of a plane boundary as in Example 10.1, we have a stepwise boundary. The discrimination on the first variable rules off three domains by hyperplanes orthogonal to that variable. The second variable rules off similarly in the region of indecision left by the first; and so on. It is possible that an optimal method based on distributions may leave a smaller residuum of undecided cases than the one we propose; but it can do so, of course, only at the expense of sacrificing the distribution-free nature of the procedure.

It may be questioned whether an improvement might be reached if we performed the operation in the reverse order, starting with the variables for which there is the greatest overlap. The point is open, but so far as these data are concerned the procedure gives much the same results.

10.28   We may add a final word on the problem of discrimination when populations differ in dispersion but not in means. It is easier to point to the problem than to suggest a solution. Consider, for example, Fig. 10.4, where the populations have the same mean but different dispersions. There are clearly areas where discrimination is possible, but the foregoing methods fail to reveal them.

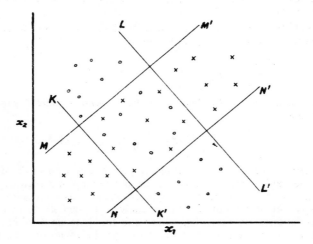

Fig. 10.4   (see text)

If the configuration was the same but the figure was rotated through 45 degrees we should arrive at meaningful results by the rank order method. The lines KK', LL' would rule off domains outside of which crosses were, so to speak, dominant; and likewise MM', NN' would define domains for the circles. The rectangle in the middle would be a zone of indecision, which would inevitably be large owing to the nature of the data.

A heuristic procedure in such cases would be to rotate the axes (for measurable variables), say, by a transformation to principal components. Lubischew (1962) has discussed the problem in a biological context. See also Bartlett and Please (1963).

10.29   One difficulty of the foregoing method stems from its sensitivity to outlying values. As we have explained it, only non-overlapping regions are accepted for discrimination; for variables of effectively infinite range there tends to be more overlap as sample size increases. It might, therefore, be preferable to accept some misclassification from the outset by permitting overlap up to a specified amount; or to fit univariate distributions and estimate the cut-off points to a specified degree of overlap. Much more remains to be done in this field.

NOTES

(1)   We have distinguished between "discrimination" and "classification". A number of earlier papers which refer to the latter in title or text are really concerned with the former.

(2)   Wald (1944) used what is effectively the linear discriminator as a classification statistic, and a good deal of theoretical work has been devoted to it. It appears to me that classification problems are in general too complicated to yield to a single statistic and that the methods of Chapter 3 are better.

(3)   Mahalanobis (see his historical note of 1948) suggests a measure of the "distance" between two populations:

$$\Delta^2 = \sum_{j,\,k=1}^{p} \alpha_{jk}(\mu_{1j} - \mu_{2j})(\mu_{1k} - \mu_{2k}),$$

where, as usual, α is the inverse of the covariance matrix. The corresponding statistic

$$D^2 = \sum_{j,\,k=1}^{p} a_{jk}(\bar{x}_{1j} - \bar{x}_{2j})(\bar{x}_{1k} - \bar{x}_{2k})$$

calculated, as concerns a_{jk}, from the pooled covariances, is of considerable theoretical interest and bears an obvious resemblance to the linear discriminator. It is not, in my opinion, suitable as a measure of the distance between points in cluster analysis. In fact, for a pair of points the covariance matrix is degenerate and the distance does not exist. To use the covariance matrix of the whole group obviously begs the question in clustering.

(4)   One of the most interesting approaches to the discrimination problem
       now made possible by the computer is the display on a television unit
       of arbitrary projections of a constellation of points on to arbitrarily
       chosen planes. With the use of a light-pen and the virtually instantan-
       eous recalculation of the necessary statistics, it is possible to explore
       the effect of removing outliers, discarding variables and so forth in a
       way which gives valuable insight into the data and the consequences
       of various ways of handling them.

(5)   For some refinements of the distribution-free method see Richards, L.E.
       (1972), *J. Roy. Statist. Soc.*, Series C, **21**, 174.

11

Categorized multivariate data

11.1   There is a large literature on methods of dealing with categorized multivariate data (CMD). Its volume alone is eloquent of the difficulties and obscurities surrounding this complex topic. Even the simplest possible case of the 2 × 2 table, the double dichotomy of attributes, can be discussed from several different points of view. In this chapter we shall try to expound the essential features of a subject which is still, in some ways, incomplete.

11.2   Typically, the kind of data with which we are concerned consists of numbers of individuals arrayed in a p-dimensional contingency table of, say, r rows, s columns, t layers, etc. We shall not often have to consider $p > 5$, and even $p = 4$ presents a problem to the draughtsman in displaying the information on a two-dimensional page. The categories for any particular variable may or may not be ordered. The data then consist of units distributed over the $rst...$ cells of the table, some of which may be empty. We assume that information in none of the cells is missing (as distinct from being known to be zero).

11.3   In one sense CMD are already clustered, and further clustering of the kind we considered in Chapter 3 is not possible without some assumptions about the nature of the categorization. If the categories are ordered and two of them, in some identifiable sense, are "close together", we may perhaps amalgamate the two corresponding arrays. As we shall emphasize later, this may give rise to misleading conclusions and is only to be recommended as a last resort when the numbers in the individual cells are small and therefore not very informative.

11.4   By the same token, transformations of the principal component type are not appropriate here. Such dependencies as exist in the polytomy have to be accepted as they stand and are not removable by rotating the axes. An attempt has been made by Lazarsfeld (1950) in his *latent structure analysis* to extend to CMD the concepts underlying factor analysis, and reference may be made to his article and a later paper by Madansky (1969). The method appears to require a rather special underlying model consisting of the admixture of a number of sets, within each of which the variables are independent; and without extraneous evidence it is usually difficult to be convinced that the model is appropriate — an echo, perhaps, of the reification problem of factor analysis itself.

11.5   Some, but not all, of the problems of CMD can be illustrated on the simple case of the 2 × 2 table in which n members are classified as possessing or not-possessing the attributes A and B:

	B	Not-B	Totals
A	a	b	$a + b$
Not-A	c	d	$c + d$
Totals	$a + c$	$b + d$	n

(11.1)

If the attributes are independent, we expect to find in the cell AB the number $(a + b)(a + c)/n$. The difference between the observed a and this "independence" frequency, say D, is given by

$$D = a - \frac{(a + b)(a + c)}{n} = \frac{ad - bc}{n}.$$

(11.2)

The "independence" table is then

$a - D$	$b + D$		$a + b$
$c + D$	$d - D$		$c + d$
$a + c$	$b + d$		n

(11.3)

The number D then summarizes the departure of the table from independence. Two quantities have been proposed as a measure of association in the table:

(1) Yule's coefficient, defined as

$$Q = \frac{ad - bc}{ad + bc}.$$

(11.4)

(2) The coefficient V, given by

$$V = \frac{ad - bc}{\{(a + b)(a + c)(b + d)(c + d)\}^{\frac{1}{2}}}.$$

(11.5)

Yule in fact considered a third coefficient,

$$\gamma = \frac{(ad)^{\frac{1}{2}} - (bc)^{\frac{1}{2}}}{(ad)^{\frac{1}{2}} + (bc)^{\frac{1}{2}}},$$

(11.6)

but it is very easy to show that

$$Q = 2\gamma/(1 + \gamma^2).$$

(11.7)

11.6 A little algebra, which we leave to the reader, shows that Q and V cannot lie outside the range -1 to 1. Both are zero when $ad - bc = 0$. In fact, the coefficient V is simply related to χ^2; for the 2×2 table,

$$\chi^2 = nV^2. \tag{11.8}$$

11.7 The three quantities Q, V or γ may be considered as measures of the intensity of association of the two attributes A and B. When we come to questions of interpretation from random samples, however, we find it necessary to distinguish three different classes of case:

(1) Both sets of marginal frequencies may be random variables. We regard the set of n as distributed over the fourfold classification at random; or, to put it another way, we take a sample of n from a bivariate population and classify into a double dichotomy.

(2) One marginal frequency may be fixed in advance. We might, for example, decide to select $a + b$ A's and $c + d$ not-A's, and then observe how they fall into B's and not-B's.

(3) Both marginal frequencies may be fixed in advance. This is a rarer case, but can occur. For example, we may select 20 men and 20 women (determining the A's and not-A's) and present each of them with two samples of butter and margarine of which *they know* that 5 are of one kind and 5 of the other, but are asked to say which are which. All the marginal totals will then be 100.

11.8 To cut short a rather long story (for which see Kendall and Stuart, vol. 2, chapter 33), a test of independence in large samples is given by χ^2 with one degree of freedom. For small samples, exact results are available, but the same test can be applied as an approximation. Its power, however, varies according to the three kinds of hypotheses delimited in 11.7. We will establish the major asymptotic result to link up the theory with the general results in hypothesis testing of Chapter 9.

In symmetrical form, which will be useful for later generalization, we may write the 2×2 table as

n_{11}	n_{12}	$n_{1.}$
n_{21}	n_{22}	$n_{2.}$
$n_{.1}$	$n_{.2}$	n

$$\tag{11.9}$$

where a dot in the suffix indicates summation. If numbers are allocated over the four cells at random, we may regard the corresponding probabilities p_{11}, p_{12}, p_{21}, p_{22} as determining a quadrinomial distribution with likelihood

$$L = \frac{n!}{n_{11}! \, n_{12}! \, n_{21}! \, n_{22}!} (p_{11})^{n_{11}} (p_{12})^{n_{12}} (p_{21})^{n_{21}} (p_{22})^{n_{22}}. \tag{11.10}$$

To estimate the p's we maximize L, subject to the constraint that $\Sigma\, p = 1$. This leads easily to the estimators

$$\hat{p}_{ij} = n_{ij}/n, \quad i, j = 1, 2, \tag{11.11}$$

which is what one might expect. It is also easy to see that these ML estimators are unbiased.

Now if we wish to test the hypothesis of independence, this is expressed as

$$p_{11} p_{22} = p_{12} p_{21}. \tag{11.12}$$

Allowing p_{11} and p_{12} to vary, and putting

$$p_{21} = \frac{p_{11}(1 - p_{11} - p_{12})}{p_{11} + p_{12}}, \quad p_{22} = \frac{p_{12}(1 - p_{11} - p_{12})}{p_{11} + p_{12}}$$

we find that the logarithm of the likelihood function, except for constants, reduces to

$$\log L = n_{.1} \log p_{11} + n_{.2} \log p_{12} + n_{2.}\{\log(1 - p_{1.}) - \log p_{1.}\}. \tag{11.13}$$

Differentiation with respect to the parameters leads to

$$\hat{p}_{11} = \frac{n_{1.}n_{.1}}{n^2}, \quad \hat{p}_{12} = \frac{n_{1.}n_{.2}}{n^2}, \tag{11.14}$$

with analogous expressions for $\hat{p}_{21}$ and $\hat{p}_{22}$.

Thus our intuitive estimate of the "independences" as frequencies from the products of the marginal proportions is justified. Substituting these estimates in (11.10), we have

$$L \propto (n_{1.}n_{.1})^{n_{11}} (n_{1.}n_{.2})^{n_{12}} (n_{.1}n_{2.})^{n_{21}} (n_{2.}n_{.2})^{n_{22}}/n^{2n}. \tag{11.15}$$

The unconditional estimates obtained by substituting from (11.11) in the likelihood function gave

$$L \propto n_{11}^{n_{11}} n_{12}^{n_{12}} n_{21}^{n_{21}} n_{22}^{n_{22}}. \tag{11.16}$$

The ratio of the likelihoods is then

$$\left(\frac{n_{1.}n_{.1}}{nn_{11}}\right)^{n_{11}} \left(\frac{n_{1.}n_{.2}}{nn_{12}}\right)^{n_{12}} \left(\frac{n_{2.}n_{.1}}{nn_{21}}\right)^{n_{21}} \left(\frac{n_{2.}n_{.2}}{nn_{22}}\right)^{n_{22}}. \tag{11.17}$$

If we write

$$l_{ij} = np_{ij} = \frac{n_{i.}n_{.j}}{n},$$

the logarithm of (11.17) may be expanded to give, as far as the second order in $n_{ij} - l_{ij}$,

$$-2 \log L = \sum_{i, j} \frac{(n_{ij} - l_{ij})^2}{l_{ij}} = \chi^2. \tag{11.18}$$

Thus, asymptotically, the hypothesis may be tested with χ^2 derived from the fourfold table. The number of degrees of freedom is unity (3 parameters in the numerator, 2 in the denominator).

11.9 Considered as measures of *intensity* of association, Q and V have very different scales, and the numerical values which they assume in particular cases may be quite different. It is not always easy to choose between them, and in cases of doubt perhaps both should be computed.

Example 11.1

The following data (from the Kenya census of 1969) show, for males and for females in the age-group 40–49 years, the cases in which father was still living or not (F or not-F) and mother was still living or not (M or not-M)

Males (000's)					Females (000's)			
	F	Not-F				F	Not-F	
M	66	106	172	M		61	94	155
Not-M	28	158	186	Not-M		27	177	204
	94	264	358			88	271	359

$$(11.19)$$

For males we find $Q = 0.557$ $V = 0.278$
For females $Q = 0.619$ $V = 0.301$.

In either case there is positive association between survival of parents, and comparison of Q for males and females, or of V for males and females, indicates that there is not much difference according to sex of respondent. But if we wish to compare with another age-group the comparison is not quite so straightforward. For example, in the age-group 20–24 the corresponding figures are

Males (000's)					Females (000's)			
	F	Not-F				F	Not-F	
M	287	86	373	M		299	90	330
Not-M	28	22	50	Not-M		31	26	116
	315	108	423			330	116	446

$$(11.20)$$

For males $Q = 0.448$ $V = 0.155$
For females $Q = 0.472$ $V = 0.198$.

Comparison with the 40–49 age-group indicates that the association is weaker in the 20–24 group, but V would indicate a greater difference than Q, and clearly we have to be careful in interpreting the relative intensities of association.

Example 11.2

Consider now a similar classification of a group of 450 patients according to sex, treatment (against non-treatment), and recovery, say A, B, C in that order. The data might, for example, be

A (Male)

	B	not-B	
C	80	100	180
Not-C	40	80	120
	120	180	300

Not-A (female)

	B	not-B	
C	20	100	120
Not-C	10	20	30
	30	120	150

$$(11.21)$$

In the male group we find $Q = 0.231$
In the female group $Q = -0.429$.

So far as these data are representative, we should conclude that for males, treatment favoured recovery, and for females the contrary.

Now if we amalgamate the two tables to group the sexes together, we find

	B	not-B	
C	100	200	300
Not-C	50	100	150
	150	300	450

$$(11.22)$$

Q is now equal to zero. We should mistakenly infer that treatment and recovery were independent.

11.10   We can illustrate the point algebraically for the $2 \times 2 \times 2$ table. Denote the two groups according to C by the suffixes 1 and 2. Then, generalizing D of (11.2),

$$D_{AB.C} + D_{AB.\text{not-}C} = a_1 - \frac{(a_1 + b_1)(a_1 + c_1)}{n_1} + a_2 - \frac{(a_2 + b_2)(a_2 + c_2)}{n_2}.$$

If letters without suffixes refer to C and not-C amalgamated, this becomes

$$a_1 - \frac{(a_1 + b_1)(a_1 + c_1)}{n_1} + a - a_1 - \frac{(a + b - a_1 - b_1)(a + c - a_1 - c_1)}{n_2}$$

$$= a - \frac{(a + b)(a + c)}{n} + \frac{(a + b)(a + c)}{n} - \frac{(a_1 + b_1)(a_1 + c_1)n}{n_1 n_2} - \frac{(a + b)(a + c)}{n_2}$$

$$+ \frac{(a + b)(a_1 + c_1)}{n_2} + \frac{(a + c)(a_1 + b_1)}{n_2}$$

$$= D_{AB} - \frac{n}{n_1 n_2}\left\{(a+b)(a+c)\frac{n_1^2}{n_2^2} + (a_1+b_1)(a_1+c_1) - (a+b)(a_1+c_1)\frac{n_1}{n}\right.$$

$$\left. - (a_1+b_1)(a+c)\frac{n_1}{n}\right\}$$

$$= D_{AB} - \frac{n}{n_1 n_2}\left\{(a_1+b_1) - (a+b)\frac{n_1}{n}\right\}\left\{(a_1+c_1) - (a+c)\frac{n_1}{n}\right\}. \qquad (11.23)$$

Now if, instead of amalgamating C and not-C we amalgamate A and not-A, n_1 is seen to be the number of C's in the resulting 2×2 table; and similarly for the amalgamation of B and not-B. Equation (11.23) then becomes

$$D_{AB.C} + D_{AB.\text{not-}C} = D_{AB} - \frac{n}{n_1 n_2} D_{AC} D_{BC}. \qquad (11.24)$$

Hence, even if A and B are independent in both C and not-C, so that the left-hand side of (11.24) vanishes, they are not independent in C and not-C combined, unless one or both of D_{AC}, D_{BC} vanish, i.e. C is independent of A or of B or of both. In a converse kind of way, even if A and B are both independent of C they are not independent of each other, even if they are independent in C alone or not-C alone.

Example 11.3

It is, in fact, possible to exhibit the effect in an even more striking form in what is sometimes known as Simpson's paradox (Simpson, 1951). Consider groups of male and female patients as follows:

Males (000's)	B	not-B		Females (000's)	B	not-B	
C	10	100	110	C	100	50	150
Not-C	100	730	830	Not-C	50	20	70
	110	830	940		150	70	220

$$(11.25)$$

Males and Females together (000's)	B	not-B	
C	110	150	260
Not-C	150	750	900
	260	900	1160

$$(11.26)$$

Taking first the group of males and females together, we see that $ad - bc$ $(110.750 - 150.150)$ is positive, so the association between treatment and recovery is positive, and on that basis one would recommend the application of the treatment. But in the male group the association is negative, and so also for the female group. Oddly, although treatment appears deleterious for either sex alone, it appears advantageous for the two together.

11.11   Effects of this kind sometimes appear in practice and cannot be dismissed as artificial. The moral appears to be that the amalgamation of classes may give rise to misleading conclusions. One might, somewhat hastily, conclude that it should never be done. Unfortunately there are two considerations, at least, which sometimes force us to compromise:

(1)   In an extensive inquiry with a considerable number of classificatory variables the number of cross-tabulations which can possibly be carried out is prohibitively large, even when a computer is available.

(2)   Continual cross-tabulation, except for enormous samples, leads to very small numbers in the sub-cells and significance tends to be lost.

The practitioner, therefore, must be prepared, in particular cases, to accept some amalgamation. But he should always be aware of its dangers and be willing to spend a good deal of time examining sub-classifications if he feels that heterogeneous material is being run together.

11.12   Consider now the two-dimensional contingency table of r rows and s columns with frequencies in the body of the table represented by n_{ij} and in the margins by $n_{i.}$ or $n_{.j}$ as the case may be. By a straightforward generalization from the 2×2 case we may calculate for each cell an "independence" frequency $n_{i.}n_{.j}/n$ and consider the rs differences

The quantity
$$d_{ij} = n_{ij} - n_{i.}n_{.j}/n. \qquad (11.27)$$

$$\chi^2 = \sum_{r,\,s} \frac{d_{ij}^2}{n_{1.}n_{.j}/n} \qquad (11.28)$$

is distributed in the standard form with $(r-1)(s-1)$ degrees of freedom. In fact, if p_{ij} is the probability that a member falls into the (i, j)th cell, the probability of the observations is

$$P\{n_{ij}|p_{ij}, n\} = \frac{n!}{\prod n_{ij}!} \prod (p_{ij})^{n_{ij}}, \qquad (11.29)$$

which we may write as

$$n! \frac{\prod_i p_{i.}^{n_{i.}}}{\prod_i n_{i.}!} \cdot n! \frac{\prod_j p_{.j}^{n_{.j}}}{\prod_j n_{.j}!} \cdot \frac{\prod_i n_{i.}! \prod_j n_{.j}!}{\prod_{i,j} n_{ij}! \, n!} \cdot \prod_{i,j} \left(\frac{p_{ij}}{p_{i.}p_{.j}}\right)^{n_{ij}}. \qquad (11.30)$$

In the case of independence, $p_{ij} = p_{i.}p_{.j}$ and the last item is unity. It is then easy to show that the three components in the product are asymptotically distributed as χ^2. The whole set of (11.26) has $rs - 1$ degrees of freedom. The row quantity

$$\chi_R^2 = \sum_i \frac{(n_{i.} - np_{i.})^2}{np_{i.}} \qquad (11.31)$$

has $r - 1$ d.f. The column quantity

$$\chi_C^2 = \sum_i \frac{(n_{.j} - np_{.j})^2}{np_{.j}} \qquad (11.32)$$

has $s - 1$ d.f. The final quantity

$$\chi^2 = \sum_{i-j} \frac{(n_{ij} - n_{i.}n_{.j}/n)^2}{n_{i.}n_{.j}/n}, \qquad (11.33)$$

has $(r - 1)(s - 1)$ d.f. The total χ^2 with $rs - 1$ d.f. may then be regarded as partitioned into three independent constituents, the d.f. summing, since

$$rs - 1 = r - 1 + s - 1 + (r - 1)(s - 1). \qquad (11.34)$$

(The independence follows from a theorem of Cochran to the effect'that if a χ^2 variable is the sum of χ^2 variables and their degrees of freedom are additive, the constituents of the sum are independent.)

11.13 When we are interested in the independence of the row and column classifications we can compute the constituent χ^2 from (11.30) to (11.33), provided that the marginal p's are given *a priori*. If not, they are estimated as $n_{i.}/n$ or $n_{.j}/n$ as the case may be, and the quantities of (11.31) and (11.32) vanish, so that we are left with the test based on (11.33) and row versus column effects are indistinguishable. This is the familiar test of elementary theory. If, however, independence is rejected, it is usually worth while examining the individual contributions to χ^2 in (11.33), just as we found it worth while considering the individual residuals in a regression. Such a scrutiny, though not conclusive, will give an idea whether the major departure from independence can be assigned to certain cells or rows or columns of the contingency table, as distinct from being an overall departure.(*)

11.14 Various measures of departure from independence have been proposed for the contingency table. Some are based on the χ^2 value of equation (11.33). That value itself is an unsuitable coefficient because it may be infinite for large n. Karl Pearson accordingly defined a *coefficient of contingency* as

$$P = \left(\frac{\chi^2}{n + \chi^2} \right)^{\frac{1}{2}}. \qquad (11.35)$$

This vanishes, as it should, when there is complete independence, and conversely it cannot exceed unity. But it cannot always attain unity. For example,

---

(*) Dr Morton B. Brown, working at the University of California, Los Angeles, has developed a program for sequentially isolating the larger contributions to a χ^2 and proceeding until the remaining part of the contingency table exhibits no significant row–column effects. See also Fienberg (1972).

in an $r \times r$ table for which all cells are empty except the main diagonal, $n_{i.} = n_{.i} = n_{ii}$ and

$$\chi^2 = n\left(\sum \frac{n_{ij}^2}{n_{i.} n_{.j}} - 1\right) = n(r-1),$$

so that

$$P = \left(\frac{r-1}{r}\right)^{\frac{1}{2}}. \tag{11.36}$$

To remedy this, Tschuprow proposed the alternative

$$T = \left[\frac{\chi^2}{n\{(r-1)(c-1)\}^{\frac{1}{2}}}\right]^{\frac{1}{2}} \tag{11.37}$$

which can attain unity in an $r \times r$ table but not for $r \times c$ if $c \neq r$.

Cramer (1946) proposed a further modification:

$$C = \left\{\frac{\chi^2}{n \min(r-1, c-1)}\right\}^{\frac{1}{2}} \tag{11.38}$$

which can always attain $+1$.

A number of other measures have been proposed by Goodman and Kruskal (1954, 1959). They are not based on χ^2 but on probabilistic considerations, e.g. on the probability of correctly predicting one variable when the value of the other variable is known.

11.15   For many purposes the interest in threefold or higher-order tables is not so much in the measure of association itself as in the way it varies according to the nature of the classification. Once again we will illustrate the general problem by reference to the simplest case, that of a $2 \times 2 \times 2$ table.

On the face of it, given three attributes, A, B and C, we can examine many hypotheses concerning their mutual relationships: for example, whether the association between A and B is the same in the class of C's as in not-C's; or whether the association between A and C is the same with respect to B and not-B; or that between B and C is the same with respect to A or not-A; or whether there is some sense in which we can say that A, B and C are associated. However, there rarely seem to be cases in which the relationships among the variables are symmetric. More usually, some of the variables are controllable and others are consequential on them; or some can be considered as stimulus variables and others as responses; or (not to become embroiled in philosophical problems of causality) some may be regarded as causes and some as effects. The division of the variables into some such categories will usually influence the way in which we analyse the data and will restrict the number of hypotheses which we wish to consider; and this is a very fortunate circumstance, because otherwise the number of such hypotheses becomes embarrassingly large.

11.16 Let p_{ijk} be the proportion of total frequency n in the ith row, jth column and kth layer. The association in the C's (corresponding to layers) is measured by the ratio

$$\frac{p_{111}\, p_{221}}{p_{121}\, p_{211}}$$

and in the not-C's by a similar expression with 2 instead of 1 in the third suffix. Thus A, B are equally associated in C and not-C if

$$\frac{p_{111}\, p_{221}}{p_{121}\, p_{211}} = \frac{p_{112}\, p_{222}}{p_{122}\, p_{212}},$$

or equivalently if

$$p_{111}\, p_{221}\, p_{122}\, p_{212} = p_{112}\, p_{222}\, p_{121}\, p_{211}. \tag{11.39}$$

This, as it happens, has a certain symmetry and could equally well be written

$$\frac{p_{111}\, p_{212}}{p_{211}\, p_{112}} = \frac{p_{211}\, p_{222}}{p_{221}\, p_{122}}, \tag{11.40}$$

which compares the association ratio of A and C in B and not-B. Also equivalent is

$$\frac{p_{111}\, p_{122}}{p_{121}\, p_{112}} = \frac{p_{211}\, p_{222}}{p_{221}\, p_{212}}, \tag{11.41}$$

comparing B and C in A and not-A. Equation (11.39), or the ratio of one side to the other, then summarizes the extent to which one pair of attributes is associated in the sub-classes determined by the third.

11.17 A situation for which (11.39) holds was described by Bartlett (1935), the first to consider it in detail, as one of *zero interaction*. The terminology was taken from the analysis of variance and, although somewhat misleading, is too firmly embedded in the literature to be disturbed. In ANOVA, when a variable y is observed at various levels of factors $f_1, f_2, \ldots, f_k$, the total sum of squares of y is partitioned into contributions from f_1, f_2, etc., separately (so-called "first-order interactions") and a residual. This residual can be partitioned into sums of squares (second-order interaction), representing the extent to which f_i and f_j are not independent; and so on. The analogy with the multivariate contingency table must not be pressed too closely (the quantities in the cells are counted numbers of occurrence, not variate-values), but it is in general accord with the idea of interaction to seek some measure of the extent to which the relationships among the variables are not independent.

Example 11.4 (Bartlett, 1935)

This example has been worked over by many later writers.

An experiment was conducted to investigate the propagation of plum root-stocks from root cuttings. There were four treatments, according to whether the cuttings were long or short and planted at once or in the spring. 240 cuttings were used for each treatment. The results were as follows:

Length	Time of planting			
of	At once		In Spring	
cutting	Alive	Dead	Alive	Dead
Long	156	84	84	156
Short	107	133	31	209

(11.42)

In this case there is one response variable (alive/dead) and two factors (long/short and time of planting).

The problem is to determine from the marginal frequencies what are the expected values in the eight cells of the table under the hypothesis of equation (11.36). Suppose that the expected value in the cell $(1, 1, 1)$ corresponding to probability p_{111} and the number n_{111} is $n_{111} + x$. Then the other expected frequencies are expressible in terms of n_{ijk} and x, and (11.36) reduces to

$$(n_{111} + x)(n_{221} + x)(n_{122} + x)(n_{212} + x)$$
$$= (n_{112} - x)(n_{222} - x)(n_{121} - x)(n_{211} - x). \qquad (11.43)$$

This is a cubic in x, in the present case giving a value of x of $5 \cdot 096$. The expected frequencies are then (displayed to correspond to (11.42)):

$$161 \cdot 096 \qquad 73 \cdot 904 \qquad 78 \cdot 964 \qquad 161 \cdot 096$$
$$101 \cdot 904 \qquad 138 \cdot 096 \qquad 36 \cdot 096 \qquad 203 \cdot 904 \qquad (11.44)$$

We can therefore calculate χ^2 as the sum of 8 terms typified by $(5 \cdot 096)^2/161 \cdot 096$. This gives a value of $2 \cdot 27$. The number of degrees of freedom is unity and the value is not significant.

The question is: not significant of what? In this case we have two experimentally controlled variables, length of cutting and time of planting. The response variable is the survival of the root-stock. For planting "at once" we have the table

	Alive	Dead	Totals
Long	156(131·5)	84(108·5)	240
Short	107(131·5)	133(108·5)	240
	263	217	480

(11.45)

The "independence frequencies" are shown in brackets. The value of χ^2 is $20 \cdot 19$ and is highly significant.

For the plantings in Spring we have

	Alive	Dead	Totals
Long	84(57·5)	156(182·5)	240
Short	31(57·5)	209(182·5)	240
Totals	115	365	480	(11.46)

χ^2 in this case is 32·02, again highly significant.

One conclusion is that, whether we plant at once or in Spring, the advantage lies with the long cuttings. The value of Q for (11.45) is 0·395, for (11.46) 0·568. Are these two significantly different? Is the beneficial survival effect of length of cutting the same for both times of planting? Or, to put it another way, is there any interaction between length of cutting and time of planting?

The answer given by our χ^2 test on (11.41) is negative. We can get some confirmation of this conclusion by considering the standard error in large samples for a value of Q, which we quote without proof from Kendall and Stuart, vol. 2, chapter 33:

$$\operatorname{var} Q = \tfrac{1}{4}(1 - Q^2)\left\{\frac{1}{a} + \frac{1}{b} + \frac{1}{c} + \frac{1}{d}\right\}^{\frac{1}{2}}. \tag{11.47}$$

In our present case, applied to (11.42) this gives a standard error of about 0·08. That of (11.43) is of the same order. The standard error of the difference of the Q's is then about $0·08\sqrt{2}, = 0·113$.

The actual difference is 0·173, less than twice the standard error.

11.18 Hypotheses of the type of (11.39) are multiplicative in character. Even the ratio for the 2×2 table, ad/bc, can be considered in that light. This consideration, together with that of symmetry, suggests that a suitable parameter to test association is, in a 2×2 table,

$$\lambda = \log \frac{p_{11} p_{22}}{p_{12} p_{21}}. \tag{11.48}$$

Example 11.5

The ratio λ in many cases is roughly equal to twice the coefficient Q, except when the latter is near unity. In fact, since

$$Q = \frac{ad - bc}{ad + bc},$$

we have

$$\frac{ad}{bc} = \frac{1 + Q}{1 - Q},$$

and hence

$$\lambda = \log ad/bc = \log(1+Q) - \log(1-Q)$$
$$= 2Q + O(Q^3). \tag{11.49}$$

For example, in the data of Example 11.1, for males in the 40–49 group,

$$Q = 0.557, \qquad \lambda = 1.257,$$

and for females in the 20–24 group,

$$Q = 0.472, \qquad \lambda = 1.025.$$

11.19   Consider next a $2 \times 2 \times t$ table and define

$$\lambda_k = \log p_{22k} - \log p_{21k} - \log p_{12k} + \log p_{11k}. \tag{11.50}$$

This has been called (Birch, 1963) the degree of partial association between rows and columns in the t-fold classification. A zero second-order interaction in such a table would correspond to the case where λ_k is the same for each layer. This clearly incorporates the $2 \times 2 \times 2$ case considered in Example 11.4.

11.20   We now begin to run into difficulties which were not apparent in dichotomous tables. Even in a two-way table, $r \times s$, there are numerous dichotomies which we can extract for separate consideration – in fact, $\frac{1}{4}r(r-1)s(s-1)$, according to which pairs of rows and columns are selected for consideration. Further complications arise when more than two dimensions are involved. In order not to get lost in the sheer number of possibilities which arise for consideration we must, in some way, systematize our approach to the analysis.

Consider, for example, an $r \times s \times t$ table. Following the type of measure of interaction (lack of independence) of (11.50) in the kth layer, we might define

$$\lambda_{ijk} = \log p_{ijk} - \log p_{i1k} - \log p_{1jk} + \log p_{11k} \tag{11.51}$$

$$= \log \frac{p_{ijk}\, p_{11k}}{p_{i1k}\, p_{1jk}}. \tag{11.52}$$

Here we have selected p_{11k} as a kind of point of reference, but other cells could be chosen; the symmetry of the 2×2 table has been lost. The hypothesis of layer interaction would then be

$$\lambda_{ijk} = \text{constant for all } k. \tag{11.53}$$

This, if obeyed, would establish independence among layers. Separate tests would be necessary for independence among rows or columns.

11.21   We may also recall the point made in Examples 11.2 and 11.3. Even if we established constant relation between rows and columns over all layers, it does not follow that the marginal row $\times$ column array, obtained by summing over all layers, bears that same relation. Suppose that (11.53) is satisfied, and let λ_{ij} be the common value of the quantities λ_{ijk} ($k = 1$, $2, \ldots, t$). Consider the marginal probabilities obtained by adding together all the t layers, denoted by p_{ij}. The corresponding function, μ_{ij}, for these marginals is

$$\mu_{ij} = \log p_{ij.} - \log p_{i1.} - \log p_{1j.} + \log p_{11.}. \qquad (11.54)$$

The point is that in general λ_{ij} is not equal to μ_{ij}. Thus, independence of row—column effects from layer to layer does not of itself enable us to make inferences about the first-order interactions in the marginal cells. We must, as Goodman (1969) and Plackett (1969) pointed out, exercise care in the order in which we test hypotheses.

11.22 One circumstance of frequent practical occurrence enables us to structure the analysis to some extent. In multivariate tables there is usually one variable which is dictated as the response variable and others as the stimulus variables (or equivalently, one as the regressand and the others as regressors). We are primarily interested in how the response varies according to the stimuli, and only secondarily in the extent to which the stimuli are related among themselves. Occasionally there will be more than one response variable; for instance, in an analysis of car accidents we may be interested in damage to human beings separately from damage to property, or even to classes of human beings such as drivers, passengers, and pedestrians. The situation is then analogous to the analysis of dependence where there are several regressands.

Example 11.6 (data from Grizzle *et al.*, 1969)

There is an unpleasant consequence which sometimes follows surgery for duodenal ulcer and is known as the "dumping syndrome". Table 11.1 shows the number of cases in four hospitals classified (1) by the severity of the condition, as none, slight, and moderate, and (2) by the severity of the operation as A (drainage and vagotomy), B (25 percent resection and vagotomy), C (50 percent resection—hemigastrectomy and vagotomy), D (75 percent resection).

This is a $3 \times 4 \times 4$ table. It is instructive to display the results, with the various marginal totals, in three dimensions but to do so imposes something of a strain on the draughtsman, as Fig. 11.1 will show. For tables in four or more dimensions, of course, complete display of this kind is impossible.

In this case we have one three-way response variable (syndrome severity) and two four-way "stimulus" variables (hospital and type of surgery). The first question we examine is whether there is any interaction between hospital (H) and type of surgery (T). We therefore examine the independence of H and T in each of the three categories of syndrome.

Table 11.1 *Hospitals classified by surgery on duodenal ulcer*

Surgical procedure	Syndrome severity	Hospital			
		1	2	3	4
A	None	23	18	8	12
	Slight	7	6	6	9
	Moderate	2	1	3	1
B	None	23	18	12	15
	Slight	10	6	4	3
	Moderate	5	2	4	2
C	None	20	13	11	14
	Slight	13	13	6	8
	Moderate	5	2	2	3
D	None	24	9	7	13
	Slight	10	15	7	6
	Moderate	6	2	4	4

In point of fact, the numbers in the "moderate" category are so small that it would be unsafe to draw any conclusions from an $H \times T$ table for "moderate" alone. We shall amalgamate "slight" and "moderate".

The frequencies for syndrome "none" are given in Table 11.2. The figures in brackets are the "independence" frequencies based on the marginal totals, e.g. in row A, column 1, $(61 \times 90)/240 = 22 \cdot 875$.

From these values we find $\chi^2 = 4 \cdot 72$ on 9 degrees of freedom. There is no evidence here of dependence.

Table 11.2 *Independence frequencies for a bivariate distribution derived from Table 11.1. (Severity of operation by type of hospital)*

	1	2	3	4	Totals
A	23 (22·875)	18 (14·742)	8 (9·658)	12 (13·725)	61
B	23 (25·500)	18 (16·433)	12 (10·767)	15 (15·300)	68
C	20 (21·750)	13 (14·017)	11 (9·138)	14 (13·050)	58
D	24 (19·875)	9 (12·808)	7 (8·392)	13 (11·925)	53
Totals	90	58	38	54	240

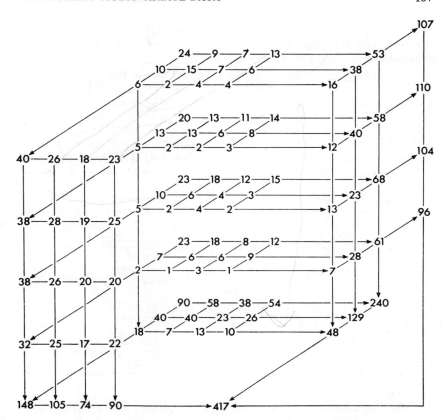

Fig. 11.1 Diagrammatic representation of the data of Table 11.1

Table 11.3 *Independence frequencies for a bivariate distribution derived from Table 11.1. (Type of hospital by treatment)*

	1	2	3	4	Totals
A	9 (11·469)	7 (9·294)	9 (7·119)	10 (7·119)	35
B	15 (11·797)	8 (9·559)	8 (7·322)	5 (7·322)	36
C	18 (17·040)	15 (13·808)	8 (10·576)	11 (10·576)	52
D	16 (17·695)	17 (14·339)	11 (10·983)	10 (10·983)	54
Totals	58	47	36	36	177

The corresponding results for $H \times T$ in the amalgamated classes "slight plus moderate" are given in Table 11.3. Here $\chi^2 = 6 \cdot 23$, also with 9 d.f., and again there is no evidence of dependence.

There being no evidence of relation between hospital and treatment, we feel justified in amalgamating the figures for the four hospitals to examine the relation between treatment and syndrome (Table 11.4).

Table 11.4   *Independence frequencies for a bivariate distribution derived from Table 11.1. (Type of treatment by severity of syndrome)*

	Syndrome None	Slight	Moderate	Totals
A	61 (55·252)	28 (29·698)	7 (11·050)	96
B	68 (59·856)	23 (32·173)	13 (11·971)	104
C	58 (63·309)	40 (34·029)	12 (12·662)	110
D	53 (61·583)	38 (33·101)	16 (12·317)	107
Totals	240	129	48	417

χ^2 for this table is $10 \cdot 54$ with 6 degrees of freedom. This is significant at the 10 percent point. We suspect that, regardless of hospital, there may be some (perhaps slender) relation between syndrome and severity of operation — not, perhaps, a medically surprising conclusion.

If necessary, we can proceed to a χ^2 test for the $3 \times 4 \times 4$ table as a whole. To save space, I omit the details which (preserving the "moderate" cells, notwithstanding the smallness of the numbers in them) give a value of χ^2 equal to $32 \cdot 471$ on 18 degrees of freedom, significant at the 4 percent point.

Grizzle, Starmer and Koch (1969), in the paper under reference, suggest constructing an index of severity of the syndrome. If the frequencies of none, slight, moderate are f_1, f_2, f_3, respectively, they propose a mean score of $(f_1 + 2f_2 + 3f_3)/(f_1 + f_2 + f_3)$. Thus, for example, the score in row A, column 1, of Table 11.1 is $(23 + 14 + 6)/32 = 1 \cdot 344$. On this basis we can perform an analysis of variance on the scores categorized by H and T. The relevant table is Table 11.5. It is convenient to simplify by reckoning from an origin of $1 \cdot 3$ and multiplying by 1000.

Table 11.5 *Scores for the data of Table 11.1 classified by type of treatment and severity of syndrome*

	1	2	3	4
A	1·344	1·320	1·706	1·500
B	1·526	1·385	1·600	1·350
C	1·605	1·607	1·526	1·560
D	1·550	1·731	1·833	1·609

The analysis of variance is

	s.s.	d.f.	s.s./d.f.	F
Between rows (Treatments)	126 594	3	42 198	3·358
Between columns (Hospitals)	75 921	3	25 307	2·014
Residual	113 095	9	12 566	
Total	315 610	15		

The effect between hospitals is not significant. That between treatments is significant at a point somewhere between 5 and 10 percent. The general conclusion is confirmed, but of course the imposition of scores to some extent begs the question.

Example 11.7 (data from Hoyt *et al.*, 1959)

Table 11.6 shows 13 968 Minnesota high-school graduates of June 1938, classified according to four factors:

R_1 to R_3: position in high-school graduating class, by thirds;
P_1 to P_4: post high-school status in April 1939, the categories being enrolled in college, enrolled in non-collegiate schools, employed full-time, and other;
L_1 to L_7: father's occupational level;
S_1 and S_2: sex.

There are several features of interest on which one might want to analyse this table. For the purpose of this example we shall examine the relation between R and P, the high-school rank and the post-high school status. We have a four-dimensional classification $3 \times 4 \times 7 \times 2$.

The two sexes and seven occupational levels give us fourteen sub-tables, each a contingency of 3×4 (6 degrees of freedom). Following Table 11.6 are given the values of χ^2 and of C, the Pearson contingency coefficient for each of the 14. (The tables themselves are omitted to save space; they can be derived from Table 11.6.)

Table 11.6　*High-school graduates classified by high-school rank, father's occupational level, sex, and scholastic performance*

Sex	Father's occu-pational level, L	R_1 lowest third				R_2 middle third				R_3 upper third			
		P_1	P_2	P_3	P_4	P_1	P_2	P_3	P_4	P_1	P_2	P_3	P_4
Male S_1	1	87	3	17	105	216	4	14	118	256	2	10	53
	2	72	6	18	209	159	14	28	227	176	8	22	95
	3	52	17	14	541	119	13	44	578	119	10	33	257
	4	88	9	14	328	158	15	36	304	144	12	20	115
	5	32	1	12	124	43	5	7	119	42	2	7	56
	6	14	2	5	148	24	6	15	131	24	2	4	61
	7	20	3	4	109	41	5	13	88	32	2	4	41
Female S_2	1	53	7	13	76	163	30	28	118	309	17	38	89
	2	36	16	11	111	116	41	53	214	225	49	68	210
	3	52	28	49	521	162	64	129	708	243	79	184	448
	4	48	18	29	191	130	47	62	305	237	57	63	219
	5	12	5	10	101	35	11	37	152	72	20	21	95
	6	9	1	15	130	19	13	22	174	42	10	19	105
	7	3	1	6	88	25	9	15	158	36	14	19	93

Values of χ^2 and C for each sex and each category of father

	Males (S_1)			Females (S_2)		
	N (number)	χ^2	C	N (number)	χ^2	C
1	885	84·28	0·295	941	76·82	0·275
2	1034	89·97	0·283	1150	48·39	0·201
3	1797	102·54	0·202	2667	196·16	0·262
4	1243	94·77	0·266	1406	86·87	0·241
5	450	19·17	0·202	571	57·72	0·303
6	438	23·89	0·231	559	40·78	0·261
7	362	25·51	0·257	467	36·97	0·271
	6209			7761		

$$(11.55)$$

The values of C do not suggest very much variation, either between sex or among father's occupation. We have worked out the χ^2 as the numbers stand, recognizing that the cell numbers are sometimes small, preferring to do so rather than amalgamate rows or columns at this stage.

If we are satisfied that the sex and occupation classes are not hetero-geneous, so far as concerns the relation between R and P, we may amalgam-ate the fourteen classes to obtain Table 11.7. For this table $\chi^2 = 1044·6$, $C = 0·263$. The result is highly significant.

Table 11.7   *Bivariate table derived from Table 11.6 (high-school rank by post-high-school status)*

	P_1	P_2	P_3	P_4	Totals
R_1	578 (1043·30)	117 (179·30)	217 (325·82)	2782 (2145·58)	3694
R_2	1410 (1577·10)	277 (271·04)	503 (492·52)	3394 (3243·34)	5584
R_3	1957 (1324·50)	284 (227·65)	512 (413·67)	1937 (2724·00)	4690
Totals	3945	678	1232	8113	13 968

$$(11.56)$$

It is now worth while examining the contributions to χ^2 from the twelve cells of the table. They are (arranged in the same order as the body of (11.56))

207·3	21·5	36·4	188·5
17·7	0·14	0·24	7·0
301·5	13·8	23·2	227·4

The contributions in the middle row are relatively small. It would appear that for the middle third of high-school graduates the distribution over the four post-graduate classes is about the average for the whole group. In the lowest group (R_1) there is a large preponderance in P_4, the "other occupation", and a large deficiency in P_1, the college-enrolled group. The converse is true of the upper group R_3. The conclusions are what we might expect.

We shall examine these data further in Example 11.8.

11.23   The necessity for proceeding with the analysis in a systematic way, even if we confine ourselves to the examination of χ^2 in marginal tables, has led writers (notably Goodman, 1969 and later papers and Plackett, 1969) to suggest the *nesting* of hypotheses; working down from greater generality towards the various possible marginals if the interaction terms are such as to justify amalgamation into marginal totals. Even so, there may be an embarrassingly large number of hypotheses to consider. Consider, for example, a three-way classification with variables denoted by A, B and C. If the univariate marginal probabilities are given *a priori*, we would test each of A, B and C against the theoretical frequencies. If the bivariate margins are given *a priori* we would test each of AB, BC, CA against the theoretical frequencies. Such cases will be exceptional, and these six would not usually arise. Most of the tests would calculate the theoretical frequencies from the observed margins, and there are 11 of them: the two-dimensional tests typified by A, B (three in number); tests in which one two-way marginal and one one-way marginal are fixed, typified by AB, C (three in number); tests in which two two-way marginals are fixed, typified by AB, AC (three in number); one in which three two-way marginals are fixed, typified by AB, AC, BC; and one in which three one-way marginals are fixed, typified by A, B, C.

For a four-way classification there are no fewer than 166 possibilities, and for more than four the numbers run into thousands.

11.24   There is therefore a strong incentive to see whether the hypotheses of interest can be parametrized in some way. If we regard the probability of occurrence in a cell as multiplicative, we can write, for example, with a three-way classification

$$\log p_{ijk} = \lambda_0 + \lambda_i^A + \lambda_j^B + \lambda_k^C + \lambda_{ij}^{AB} + \lambda_{jk}^{BC} + \lambda_{ik}^{AC} + \lambda_{ijk}^{ABC}, \quad (11.57)$$

where the summation of any λ over any suffix is zero. The resemblance to an analysis-of-variance model is obvious. If the suffix i refers to the A classification, etc., the term λ_i^A will represent the contribution of A to the ith row of the table; similarly λ_{ij}^{AB} will represent an interaction term giving the contribution of A and B to the ith row and jth column, other than what is contributed by the main effects λ_i^A and λ_j^B; and so on. If λ^{AB} and subsequent terms are zero, the variables are independent.

11.25   The model of (11.57) can be regarded as an extension of a method used in the analysis of binary response data (Cox, 1970). With a single categorization we have

$$\log p_i = \lambda_0 + \lambda_i$$

$$\log \frac{p_i}{1 - p_i} = \frac{\lambda_0 + \lambda_i}{1 - \lambda_0 + \lambda_i}. \quad (11.58)$$

11.26   If a model of type (11.57) is fitted, a test of its adequacy can be carried out by χ^2 of the usual kind:

$$\chi^2 = \sum \frac{(O - T)^2}{T}, \quad (11.59)$$

where O is the observed and T the theoretical frequency. It is also possible to test by the likelihood-ratio statistic:

$$\chi_{ML}^2 = \sum \left(-2O \log \frac{O}{T} \right), \quad (11.60)$$

where T is the maximum-likelihood estimate of the cell frequencies. Both are distributed asymptotically as χ^2 with $n - p$ degrees of freedom, where n is the number of cells in the table and p is the number of parameters fitted by the model. In arithmetic practice there does not seem to be much difference, but as Brown (1973) points out, the χ^2_{ML} statistic has the advantage of being additive under partitioning; that is to say, if X_1 and X_2 are two models such that the marginals fitted by X_1 are a subset of those fitted by X_2, then

$$\chi^2_{\mathrm{ML}}(X_1) = \chi^2_{\mathrm{ML}}(X_1|X_2) + \chi^2_{\mathrm{ML}}(X_2). \tag{11.61}$$

Example 11.8

I am indebted to Dr Morton Brown for an analysis of the data of Example 11.7 on a program which he wrote at the University of California, Los Angeles.

Here we have a fourfold classification with $(4 \times 3 \times 7 \times 2) - 1 = 167$ degrees of freedom. The model is of the type of equation (11.57) carried to the fourth-order term.

Effect	d.f.		χ^2_{ML}
P	3		9851·57
R	2		388·82
L	6		4547·88
S	1	12	172·77
PR	6		1062·12
PL	18		1422·78
PS	3		361·91
RL	12		172·25
RS	2		387·35
LS	6	47	58·19
PRL	36		51·29
PRS	6		3·27
PLS	18		46·80
RLS	12	72	19·28
PRLS	36	36	45·13
		167	

The first four entries in the table are, in this case, trivial. All the values of χ^2 are highly significant, but they only test the hypothesis that the frequencies in the univariate margins are all equal.

The second group tests the bivariate margins. For example, the line corresponding to *PR* tests the two-way table obtained by summing over the unspecified variables *L* and *S*, namely the relation between high-school graduation *R* and post-employment *P*. The value of Pearson's χ^2 found in Example 11.7, namely 1044·6, differs negligibly from the ML value of 1062·1.

From the other bivariate marginals, which are all significant, the situation appears to be fairly complicated. There is even an association *LS* between sex of student and father's occupation. Let us look at the primary data.

	L_1	L_2	L_3	L_4	L_5	L_6	L_7	Totals
S_1	885	1034	1797	1243	450	436	362	6207
S_2	941	1150	2667	1406	571	559	467	7761
Totals	1826	2184	4464	2649	1021	995	829	13 968

There is a large disparity in group L_3, as noticed by Ku, Varner and Kullback (1971), which contributes 41 to a χ^2 of 58 and raises some doubts about the accuracy of the data. Failing further inquiries into this particular set of data, we remove L_3 and re-run on the remaining six classes of father's occupation:

Effect	d.f.	χ^2_{ML}
PR	6	723·33
PL	15	954·20
PS	3	221·78
RL	10	144·92
RS	2	308·87
LS	5	11·36
PRL	30	29·01
PRS	6	1·07
PLS	15	25·51
RLS	10	13·96
PRLS	30	39·92
	132	

LS is not significant at the 5 percent point, but we may perhaps accept it.

The other two-way tables are significant, but the third-order terms such as PRL are not, except perhaps PLS which is significant at 5 percent. The fourth-order term also is not significant.

We may then conclude that the model is adequately represented by (11.57) with the one-way and two-way terms retained but the others omitted. Apart from LS, the two-way marginals are significant. Let us look at RL and RS.

	L_1	L_2	L_4	L_5	L_6	L_7	
R_1	361	479	725	297	324	234	2420
	(464·95)	(556·11)	(674·51)	(259·98)	(253·36)	(211·09)	
R_2	691	852	1057	409	404	354	3767
	(723·75)	(865·65)	(1049·96)	(404·68)	(394·38)	(328·58)	
R_3	774	853	867	315	267	241	3317
	(637·29)	(762·24)	(924·53)	(356·34)	(347·27)	(289·33)	
	1826	2184	2649	1021	995	829	9504

$$(11.62)$$

	S_1	S_2	
R_1	1430 (1122·92)	990 (1297·08)	2420
R_2	1790 (1747·95)	1977 (2019·05)	3767
R_3	1190 (1539·13)	2127 (1777·87)	3317
	4410	5094	9504

$$(11.63)$$

In (11.62) the expected and observed values in row R_2 are roughly equal. In R_3 there are more than expected in the L_1 and L_2 classes and a corresponding deficiency in the remainder; and conversely in R_1. It appears that better grades at high school are obtained by children of the higher occupation classes. In (11.63) there are more males than expected in the lowest grades of high school and correspondingly fewer in the highest grades.

The results for RL, RS and LS are not directly relevant to our inquiry, which was concerned mainly with the relation between P and R. Nevertheless they have an important bearing on our interpretation of that relationship. The situation is similar to that in regression analysis, where correlations among the regressands may affect our interpretation of particular terms in the regression equation. The situation in our present example seems to be (apart from the anomalous L_3 group) that sex, father's occupation and high-school results are not independent and are jointly acting on the P variable. As for regression, it is not possible without further information to disentangle their individual effects.

NOTES

(1) For a more detailed study of the problems considered in this chapter reference may be made to the monograph by Plackett (1974) and a series of papers by Goodman (1970, 1971a, 1971b, 1972a, 1972b). See also the survey by Plackett (1969).

(2) A different approach from the point of view of information theory has been developed by Kullback and colleagues (Kullback *et al.*, 1962; Ku and Kullback, 1968).

(3) For some work on the effect of collapsing multivariate tables see Bishop (1971), and for some work on incomplete tables see Fienberg (1972).

(4) Published data for three-way tables or those of higher dimension being rather scarce, certain examples have been worked over by many different authors and it is interesting to compare their conclusions, which are not always identical. We may refer particularly to some market research data by Ries and Smith (1963), some biological data by Kastenbaum and Lamphiear (1959), and the Minnesota students data discussed in the text.

Appendix A

Note on multidimensional scaling

A 1 Although, in some situations, we have to force a set of objects or individuals into an order which by its nature is one-dimensional, there is often a feeling that the enforcement is too drastic, and that the objects are really being considered in more than one dimension. Degrees of "similarity", preferences, decisions, and comparisons between groups are all of this kind. When the n objects are measured on p variables we can, by one of the techniques considered in earlier chapters, discuss the essential dimensionality of the set and, perhaps in an approximative sense, reduce the dimension to a more convenient number. But this involves measurement, and the question arises whether we can do something of the same kind in a non-metrical sense.

A 2 Let us suppose that we have a distance-metric of some kind, perhaps of the type we used in Chapter 3 on cluster analysis, perhaps based on similarity indices, perhaps based on purely subjective judgement. We imagine this given for the $\frac{1}{2}n(n-1)$ pairs of points by a number δ_{ij}, the distance or dissimilarity between the ith and jth objects.

Now we seek, in a space of lower dimensions $t < p$, another set of n points with distances say d_{ij}. If we can so choose them that the $\frac{1}{2}n(n-1)$ numbers d_{ij} have the same ranking order as the δ_{ij}, we may regard the co-ordinate values in t dimensions as accurately reproducing the pattern of dissimilarity in the original p dimensions in the sense that the distances between pairs of objects occur in the same order. The lowest value of t for which this is possible may then be regarded as a basic irreducible dimension number. Scaling or measurement on a lower number is bound to violate the data to some extent.

A 3 The procedure is only non-metric in a limited sense. To apply the idea in practice, we must have a sufficiently explicit numerical measure of dissimilarity or distance to determine the numbers δ_{ij} or at least their rank-order; and so for d_{ij}, although in the latter case we determine the values in some optimized sense such as a least-squares fit.

A 4   In practice, of course, perfect agreement between δ_{ij} and d_{ij} will be exceptional unless $t = p$. As in the case of rejecting components or redundant variables, we must be prepared to tolerate some difference. The first problem is to set up a measure of divergence between the two. Kruskal (1964a) proposed the following.

We select a number t and consider n points in it, with distances d_{ij}. (Where we put the points will be considered in a moment.) Then the *stress S* of the configuration is defined as

$$S^2 = \frac{\sum_{i<j} (d_{ij} - \hat{d}_{ij})^2}{\sum_{i<j} d_{ij}^2} , \tag{A.1}$$

where the $\hat{d}_{ij}$ are chosen so as to minimize S, subject to the constraint that they have the same order as the δ_{ij}. The denominator in (A.1) is an attempt to standardize. If all the distances d are multiplied by the same constant, the stress is unaltered; but this is not so if some of the t dimension measure-numbers are multiplied by different constants.

A 5   The "stress" is then a measure of the extent to which we are forcing the situation into lower dimensions. How much stress we are prepared to tolerate is a matter for arbitrary decision. Kruskal regards anything greater than 20 percent as poor, 10 percent as fair, 5 percent or less as good. These are not significance points in the ordinary probabilistic sense of classical statistics.

A 6   The determination of the points in the t-space and the associated d's and $\hat{d}$'s is a matter for an iterative computer routine which Kruskal describes in 1964b. A convenient set of points is chosen in the t-space as a "blast-off" basis from which to start. The distances d are calculated. Some computational procedure such as the method of steepest descent is applied to estimate the $\hat{d}$'s and to compute S. This is then reduced by moving the points, and so on.

A 7   Strictly speaking, in looking for the appropriate value of t we should follow this procedure for a set of trial values of t and consider the minimal stress for each. This involves a lot of computation and in practice, apparently, one stops short at a reasonably low value of t.

REFERENCES

A comprehensive bibliography of Multivariate Statistics by T.W. Anderson, S. Das Gupta and G.P.H. Styan has been published (1972), Oliver and Boyd, Edinburgh. The following references are confined to a few books and papers and those articles specifically mentioned in the text.

Adelman, Irma; Grier, Marsla; and Morris, Cynthia T. (1969), Instruments and goals in economic development. *Am. Econ. Rev.*, **59**, 409.

Afifi, A.A., and Elashoff, R.M. (1966, 1967), Missing observations in multivariate statistics. I. Review of literature. *J. Am. Statist. Assoc.*, **61**, 595. II. Point estimation in simple linear regression. *J. Am. Statist. Assoc.*, **62**, 10.

Anderson, T.W. (1958), *Introduction to Multivariate Statistical Analysis*. Wiley, New York.

Andrews, D. (1972), Plots of high-dimensional data. *Biometrics,* **28**, 125.

Banks, C.H. (1954), The factorial analysis of crop productivity. *J. Roy. Statist. Soc.*, **B**, **16**, 100.

Barnard, M.M. (1935), The secular variations of skull characters in four series of Egyptian skulls. *Ann. Eugen. Lond.*, **6**, 352.

Bartlett, M.S. (1935), Contingency table interactions. *J. Roy. Statist. Soc.*, *Supp.*, **2**, 248.

Bartlett, M.S. (1947), Multivariate analysis. *J. Roy. Statist. Soc.*, **B**, **9**, 176.

Bartlett, M.S. (1948), Internal and external factor analysis. *Brit. J. Psychol. (Stat. Sect.)*, **1**, 73.

Bartlett, M.S. (1949), Fitting a straight line when both variables are subject to errors. *Biometrics*, **5**, 207.

Bartlett, M.S. (1950), Tests of significance in factor analysis. *Brit. J. Psychol. (Stat. Sect.)*, **3**, 77.

Bartlett, M.S. (1951a), A further note on tests of significance in factor analysis. *Brit. J. Psychol. (Stat. Sect.)*, **4**, 1.

Bartlett, M.S. (1951b), The effect of standardization on an approximation in factor analysis. *Biometrika*, **38**, 337.

Bartlett, M.S. (1954), A note on the multiplying factors for various approximations. *J. Roy. Statist. Soc.*, **B**, **16**, 296.

Bartlett, M.S. and Please, N.W. (1965), Discrimination in the case of zero mean differences. *Biometrika*, **50**, 17.

Beale, E.M.L. (1969), Computational Methods for Least Squares. *Integer and Nonlinear Programming* (ed. J. Abadie), North-Holland Publishing Co., Amsterdam.

Beale, E.M.L., Kendall, M.G. and Mann, D.W. (1967), The discarding of variables in multivariate analysis. *Biometrika*, **54**, 357.

Beale, E.M.L. and Little, R.J.A. (1975), Missing values in multivariate analysis. *J. Roy. Statist. Soc.*, **B**, **37**, 129.

Bellman, R., and Roth, R. (1969), Curve fitting by segmented straight lines. *J. Am. Statist. Assoc.*, **64**, 1079.

Berkson, J. (1950), Are there two regressions? *J. Am. Statist. Assoc.*, **47**, 164.

Bhapkar, V.P. and Koch, G.G. (1968), On the hypothesis of no interaction in contingency tables. *Technometrics*, **10**, 107.

Birch, M.W. (1963), Maximum likelihood in three-way contingency tables. *J. Roy. Statist. Soc.*, **B**, **25**, 220.

Bishop, Y.M.M. (1971), Effects of collapsing multidimensional contingency tables. *Biometrics*, **27**, 545.

Blackith, R.E. (1960), A synthesis of multivariate techniques to distinguish patterns of growth in grasshoppers. *Biometrika*, **16**, 28.

Bolshev, L.N. (1970), Cluster Analysis. *Proc. Int. Statist. Inst.*, London meeting.

Brown, M.B. (1973), Aids in the selection of models for multidimensional contingency tables. Available from Health Sciences Computing Facility, University of California, Los Angeles.

Brown, R.L. (1957), Bivariate structural relation. *Biometrika*, **44**, 84.

Brown, R.L. and Fereday, F. (1958), Multivariate linear structural relations. *Biometrika*, **45**, 136.

Buck, S.F. (1960), A method of estimation of missing values in multivariate data suitable for use with an electronic computer. *J. Roy. Statist. Soc.*, **B**, **22**, 302.

Cochran, W.G. (1964), On the performance of the linear discriminant function. *Technometrics*, **6**, 169.

Cochran, W.G. and Bliss, C.I. (1948), Discriminant functions with covariance. *Ann. Math. Statist.*, **19**, 151.

Cochran, W.G. and Hopkins, C.E. (1961), Some classification problems with multivariate quantitative data. *Biometrics*, **17**, 10.

Cormack, R.M. (1971), A review of classification. *J.R. Statist. Soc.*, **A**, **134**, 321.

Cox, D.R. (1970), *Analysis of Binary Data*. Methuen, London.

Cramer, H. (1946), *Mathematical Methods of Statistics*. Princeton University Press.

Darroch, J.N. (1962), Interactions in multi-factor contingency tables. *J. Roy. Statist. Soc.*, **B**, **24**, 251.

Draper, N.R. (1963), "Ridge analysis" of response surfaces. *Technometrics*, **5**, 469.

Draper, N.R., and Smith, H. (1970), Methods for selecting variables from a given set of variables for regression analysis. *ISI (London) Conference Proceedings*.

Durbin, J. (1954), Errors in variables. *Rev. Int. Statist. Inst.*, **22**, 23.

Durbin, J. and Watson, G.W. (1971), Testing for serial correlation in least squares regression. *Biometrika*, **58**, 1.

Elashoff, J.E., Elashoff, R.M. and Goldman, G.E. (1967), On the choice of variables in classification problems with dichotomous variables. *Biometrika*, **54**, 668.

Fienberg, S.E. (1972), The analysis of incomplete multivariate contingency tables. *Biometrics*, **28**, 177.

Fisher, F.M. (1957), *The Identification Problem in Econometrics*, McGraw-Hill, New York.

Fisher, R.A. (1936), The use of multiple measurements in taxonomic problems. *Ann. Eugen. Lond.*, **7**, 179.

Fisher, R.A. (1938), The statistical utilization of multiple measurements. *Ann. Eugen. Lond.*, **8**, 376.

Fisk, P.R. (1967), *Stochastically Dependent Equations*. Charles Griffin & Co., London & High Wycombe.

Garside, M.J. (1965), The best sub-set in multiple regression analysis. *Applied Statistics*, **14**, 196.

Geary, R.C. (1949), Determination of linear relations between systematic parts of variables with errors of observation the variances of which are unknown. *Econometrica*, **17**, 30.

Geary, R.C. (1953), Non-linear functional relationship between two variables when one variable is controlled. *J. Am. Statist. Assoc.*, **48**, 94.

Geary, R.C. (1970), Relative efficiency of count of sign changes for assessing residual autoregression in least-squares regression. *Biometrika*, **57**, 123.

Geary, R.C. (1972), Two exercises in simple regression. *Economic and Social Review*, **3**, 551.

Glahn, H.R. (1968), Canonical correlation and its relationship to discriminant analysis and multiple regression. *J. Atmospheric Sciences*, **25**, 23.

Good, I.J. (1963), Maximum entropy for hypothesis formulation, especially for multidimensional contingency tables. *Ann. Math. Statist.*, **34**, 911.

Goodman, L.A. (1964a), Interactions in multidimensional contingency tables. *Ann. Math. Statist.*, **35**, 632.

Goodman, L.A. (1964b), Simple methods for analyzing three-factor interactions in contingency tables. *J. Am. Statist. Assoc.*, **59**, 519.

Goodman, L.A. (1968), The analysis of cross-classified data: independence, quasi-independence and interaction in contingency tables with or without missing entries. *J. Am. Statist. Assoc.*, **63**, 1091.

Goodman, L.A. (1969), On partitioning and detecting partial association in three-way contingency tables. *J. Roy. Statist. Soc*, **B, 31**, 486.

Goodman, L.A. (1970), The multivariate analysis of quantitative data: interactions among multiple classifications. *J. Am. Statist. Assoc.*, **65**, 226.

Goodman, L.A. (1971a), The analysis of multidimensional contingency tables: stepwise procedures and direct estimation methods for building models for multiple classifications. *Technometrics*, **13**, 33.

Goodman, L.A. (1971b), Partitioning of chi-square, analysis of marginal contingency tables and estimation of expected frequencies in multidimensional contingency tables. *J. Am. Statist Assoc.*, **66**, 339.

Goodman, L.A. (1972a), A general model for the analysis of surveys. *Am. J. Sociology*, **77**, 1035.

Goodman, L.A. (1972b), A modified multiple regression approach to the analysis of dichotomous variables. *Am. Soc. Rev.*, **37**, 28.

Goodman, L.A. and Kruskal, W. (1954, 1959, 1963), Measures of association for cross-classifications: Parts 1, 2, 3. *J. Am. Statist. Assoc.*, **49**, 732.

Grizzle, J.E., Starmer, C.F. and Koch, G.G. (1969), Analysis of categorical data by linear models. *Biometrics*, **25**, 489.

Haitovsky, Y. (1968), Missing data in regression analysis. *J. Roy. Statist. Soc.*, **B, 30**, 67.

Hannan, E.H. (1972), *Multiple Time-series*. Wiley, New York.

Hills, M. (1966), Allocation rules and their error rates. *J. Roy. Statist. Soc.*, **B, 28**, 1.

Hills, M. (1967), Discrimination and allocation with discrete data. *Applied Statist.*, **16**, 237.

Hotelling, H. (1931), The generalization of Student's ratio. *Ann. Math. Statist.*, **2**, 360.

Hotelling, H. (1936), Relations between two sets of variates. *Biometrika*, **28**, 321.

Hoyt, C.J., Krishnaiah, P.R. and Torrance, E.P. (1959), Analysis of complex contingency data. *J. Exp. Educ.*, **27**, 187.

Hudson, D.J. (1966), Fitting segmented curves whose join points have to be estimated. *J. Am. Statist. Assoc.*, **61**, 1097.

Jackson, P.H., Novick, M.R. and Thayer, D.T. (1971), Estimating regressions in *m* groups. *Brit. J. Math. Statist. Psychol.*, **24**, 129.

Joliffe, I.T. (1972), Discarding variables in a principal component analysis: I. Artificial data. *J. Roy. Statist. Soc.*, **C, 21**, 160.

Jöreskog, K.G. (1967), Some contributions to maximum likelihood factor analysis. *Psychometrika*, **32**, 165.

Jöreskog, K.G. (1969), A general approach to confirmatory maximum likelihood factor analysis. *Psychometrika*, **34**, 183.

Jöreskog, K.G. (1971), A general method for analysis of covariance structures. *Biometrika*, **57**, 239.

Kaiser, H.F. (1958), The varimax criterion for analytic rotation in factor analysis. *Psychometrika*, **23**, 187.

Kastenbaum, M.A. and Lamphiear, D.E. (1959), Calculation of chi-square to test the no three-factor hypothesis. *Biometrics*, **15**, 107.

Kendall, M.G. (1939), The geographical distribution of crop productivity in England. *J. Roy. Statist. Soc.*, **102**, 21.

Kendall, M.G. (1971), *Rank Correlation Methods* (4th edn). Charles Griffin & Co., London and High Wycombe.

Kendall, M.G. (1973), *Time-Series*. Charles Griffin & Co., London and High Wycombe.

Kendall, M.G. and O'Muircheataigh, C. (1977), Path analysis. World Fertility Survey Technical Paper No. 414, W.F.S., 35 Grosvenor Gardens, London, SW1W 0BS.

Kendall, M.G. (1979), The misuse of dummy variables in regression analysis. World Fertility Survey Technical Paper No. 1141, W.F.S., 35 Grosvenor Gardens, London, SW1W 0BS.

Kendall, M.G. and Stuart A. (1958 onwards), *The Advanced Theory of Statistics*, 3 vol. Charles Griffin & Co., London and High Wycombe. (Vol. 1, 4th edn, 1977; vol. 2, 4th edn, 1979; vol. 3, 3rd edn, 1976).

Kruskal, J.B. (1964a), Multidimensional scaling by optimizing goodness-of-fit to a non-metric hypothesis. *Psychometrika*, **29**, 1.

Kruskal, J.B. (1964b), Non-metric multidimensional scaling: a numerical method. *Psychometrika*, **29**, 115.

Ku, H.H. and Kullback, S. (1968), Interaction in multidimensional contingency tables: an information theoretic approach. *J. Res. Nat. Bur. Stds*, **B**, **72**, 159.

Ku, H.H., Varner, R.N. and Kullback, S. (1971), Analysis of multidimensional contingency tables. *J. Am. Statist. Assoc.*, **66**, 55.

Kullback, S., Kupperman, M. and Ku. H.H. (1962), Tests for contingency tables and Markov chains. *Technometrics*, **4**, 573.

Kullback, S. and Rosenblatt, H.M. (1957), On the analysis of multiple regression in *K* categories. *Biometrika*, **44**, 67.

Lancaster, H.O. (1969), Contingency tables of higher dimensions. *Bull. Int. Statist. Inst.*, **43**, 143.

Lawley, D.N. (1956), Tests of significance for the latent roots of covariance and correlation matrices. *Biometrika*, **43**, 128.

Lawley, D.N. and Maxwell, A.E. (1971), *Factor Analysis as a Statistical Method*. Butterworths, London. (First edition 1963. Second edition of 1971 very largely re-written.)

Lazarsfeld, P.F. (1950), The logical and mathematical foundations of latent structures analysis, in *Measurement and Prediction* (ed. Stouffer), Princeton University Press, 362.

Lederman, W. (1937), On the rank of the reduced correlational matrix in multiple factor analysis. *Psychometrika*, **2**, 83.

Lindley, D.V. (1947), Regression lines and linear functional relationship. *Supp. J. Roy. Statist. Soc.*, **9**, 218.

Linhart, H. (1959), Technique for discriminant analysis with discrete variables. *Metrika*, **2**, 138.

Longley, J.W. (1967), An appraisal of least-squares programs for the electronic computer from the point of view of the user. *J. Am. Statist. Assoc.*, **62**, 819.

Lubischew, A.A. (1962), On the use of discriminant functions in taxonomy. *Biometrics*, **18**, 455.

McGee, V.E. (1965), Invariance of personal characteristics of voice over two vowel sounds. *Perceptual and Motor Skills*, **21**, 529.

McGee, V.E. and Carleton W.T. (1970), Piecewise regression. *J. Am. Statist. Assoc.*, **65**, 1109.

Madansky, A. (1969), Latent Structure. *Ency. Social Sci.*

Mahalanobis, P.C. (1948), Historical note on the D^2-statistic. *Sankhyā*, **9**, 237.

Mann, H.B. and Wald, A. (1943), On the statistical treatment of linear stochastic equations. *Econometrica*, **11**, 173.

Moser, C.A. and Scott, W. (1961), *British Towns*. Oliver and Boyd, Edinburgh.

Mosteller, F. (1968), Association and estimation in contingency tables. *J. Am. Statist. Assoc.*, **63**, 1.

Nair, K.R. and Shrivastava, M.P. (1942), On a simple method of curve fitting. *Sankhyā*, **6**, 121.

Oldham, P.D. and Rossiter, C.E. (1965), Mentality in Coalworkers' Pneumoconiosis related to lung function: a prospective study. *Brit. J. Ind. Medicine*, **22**, 93.

Oosterhoff, J. (1963), On the selection of independent variables in a regression equation. *Mathematisch Centrum, Amsterdam Report* S 319 (VP 23).

Orchard, T. and Woodbury, M.A. (1972), A missing information principle: theory and applications. *Proc. Sixth Berkeley Symp. Math. Statist. and Prob.*, vol. I, 697.

Pearson, E.S. and Wilks, S.S. (1933), Methods of statistical analysis appropriate for K samples of two variables. *Biometrika*, **25**, 353.

Penrose, L.S. (1947), Some notes on discrimination. *Ann. Eugen. Lond.*, **13**, 228.

Phillips, A.W. (1958), The relation between unemployment and the rate of change of money wage rates in the United Kingdom, 1861–1913. *Economica*, **25**, 283.

Plackett, R.L. (1969), Multidimensional contingency tables: a survey of models and methods. *Bull. Int. Statist. Inst.*, **43** (1), 133.

Plackett, R.L. (1974), *The Analysis of Categorical Data*. Charles Griffin & Co., London & High Wycombe.

Quandt, R.E. (1960), Tests of the hypothesis that a linear regression system obeys two separate regimes. *J. Am. Statist. Assoc.*, **55**, 324.

Quenouille, M.H. (1957, 1968), *The Analysis of Multiple Time-series*. Charles Griffin & Co., London & High Wycombe.

Rao, C.R. and Slater, P. (1949), Multivariate analysis applied to differences between neurotic groups. *Brit. J. Psychol.*, **2**, 17.

Rao, C.R. (1969), *Advanced Statistical Methods in Biometric Research*. Wiley, New York.

Rhodes, E.C. (1937), An index of business activity. *J.R. Statist. Soc.*, **A**, **100**, 18.

Ries, P.N. and Smith, H. (1963), The use of chi-square for preference testing in multidimensional problems. *Chemical Engineering Progress, Symposium Series*, **59**, 39.

Robinson, D.E. (1964), Estimates for the points of intersection of two polynomial regressions. *J. Am. Statist. Assoc.*, **59**, 214.

Shepherd, R.N. (1957), Stimulus and response generalization: a stochastic model relating generalization to distance in psychological space. *Psychometrika*, **22**, 325.

Shepherd, R.N. (1962), The analysis of proximities, multidimensional scaling with an unknown distance function. *Psychometrika*, **27**, 125 and 219.

Simpson, C.H. (1951), The interpretation of interactions in contingency tables. *J.R. Statist. Soc.*, **B**, **13**, 238.

Slater, P., Shields, J. and Slater, Eliot (1964), Quadratic discriminant of zygosity from fingerprints. *J. Med. Genetics*, **1**, 42.

Smith, Cedric A.B. (1947), Some examples of discrimination. *Ann. Eugen. Lond.*, **13**, 272.

Solari, M.E. (1969), The "maximum likelihood solution" of the problem of estimating a linear functional relationship. *J.R. Statist. Soc*, **B**, **31**, 372.

Spearman, C. (1904), The proof and measurement of association between two things. *Am. J. Psychol.*, **15**, 72 and 202.

Stamp, D. (1952), *Land for Tomorrow*. Bloomington Press, Indiana.

Stone, J.R.N. (1945), The analysis of market demand. *J. Roy. Statist. Soc.*, **108**, 308.

Stone, J.R.N. (1947), The interdependence of blocks of transactions. *J. Roy. Statist. Soc., Supp.*, **9**, 1.

Theil, H. (1950), A rank-invariant method of linear and polynomial regression analysis. *Indag. Math.*, **12**, 85 and 173.

Thomson, Sir Godfrey (1939), *The Factorial Analysis of Human Ability*. London U.P.

Vinod, H.D. (1975), An extension of the Durbin–Watson test. (In course of publication.)

Wald, A. (1944), On a statistical problem arising in the classification of an individual into one of two groups. *Ann. Math. Statist.*, **15**, 145.

Wampler, R.H. (1969), Evolution of least squares computer programs. *J. Res. Nat. Bur. Stds*, **B**, **73**, No. 2, 59.

Waugh, F.W. (1942), Regression between sets of variates. *Econometrica*, **10**, 290.

Wilks, S.S. (1932), Certain generalizations in the analysis of variance. *Biometrika*, **24**, 471.

Wilks, S.S. (1935), On the independence of K sets of normally distributed statistical variables. *Econometrica*, **3**, 309.

Wilks, S.S. (1946), Sample criteria for testing equality of means, etc., in a normal multivariate system. *Ann. Math. Statist.*, **17**, 257.

Williams, R.J. (1952), Use of scores for the analysis of association in contingency tables. *Biometrika*, **39**, 274.

AUTHOR INDEX

207

SUBJECT INDEX